1L 49.95
65B

Variable stars are not of constant luminosity, and their variations may be cyclical, aperiodic, sporadic, or completely unpredictable, depending on type. The study of the variations and of the different kinds of variable star contributes to our understanding of star types, stellar evolution, and the cosmic distance scale. The measurement of variable star properties is an important field in which amateur astronomers can contribute primary research data. This book presents results from the first European meeting of the American Association of Variable Star Observers (Brussels, July 1990) which was attended by professional and amateur astronomers from 30 countries.

The 38 contributions in this book are all by professional astronomers, each of whom has a special understanding of the work done by amateurs. The chapters are aimed at researchers, advanced amateurs, and students. The authors cover such topics as the history of variable star research, the contributions of amateurs, current understanding of the major classes of variable stars, and prospects for amateur-professional collaboration in the future. Visual, photographic, photoelectric and CCD observing techniques are discussed in a practical way. Throughout the book there are many suggestions for research projects that can be undertaken with modest equipment under average observing conditions. The book conveys the tremendous enthusiasm, dedication, and skill of variable star observers worldwide, and the many benefits of international co-operation.

Variable star research: an international perspective

Variable Star Research: An International Perspective

Proceedings of the first European meeting of the American Association of Variable Star Observers: "International Cooperation and Coordination in Variable Star Research

Brussels, Belgium
24–28 July, 1990

Edited by

John R. Percy,
University of Toronto

Janet Akyuz Mattei
American Association of Variable Star Observers

Christiaan Sterken
Vrije Universiteit Brussel

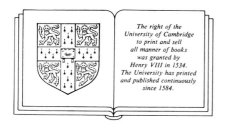

CAMBRIDGE UNIVERSITY PRESS

Cambridge

New York   Port Chester

Melbourne   Sydney

Published by the Press Syndicate of the University of Cambridge
The Pitt Building, Trumpington Street, Cambridge CB2 1RP
40 West 20th Street, New York, NY 10022–4211, USA
10 Stamford Road, Oakleigh, Victoria 3166, Australia

© Cambridge University Press 1992

First published 1992

Printed in Great Britain at the University Press, Cambridge

British Library cataloguing in publication data available

Library of Congress cataloguing in publication data available

ISBN 0 521 40469 X hardback

```
QB
833
.V37
1992
```

CONTENTS

Page

Preface xi
Group photograph xiii
List of participants xiv

Prologue: International Cooperation and Coordination 1

International Cooperation and Coordination in Astronomical Research
    Adriaan Blaauw 3

1. History and Organization   9

Variable Stars: A Historical Perspective
    John R. Percy 11
Some Little Known Variable Star Astronomers from Around the World
    Thomas R. Williams 21
AAVSO and Variable Star Observing
    Janet Akyuz Mattei 36
International Astronomical Union Commission 27: Variable Stars
    Michel Breger 50
The Present State of the Compilation of the General Catalogue of Variable Stars
    N. Samus 52
The Extragalactic Variables in the Fourth Edition of the GCVS
    N.A. Lipunova 55
A Catalogue of Variable Components in Visual Double and Multiple Stars
    P. Lampens 60

2. Visual Observation   65

Coordination of Visual Observing Programs
    Janet Akyuz Mattei 67
On the Homogeneity of Visual Photometry
    C. Sterken and J. Manfroid 75
Visual Searching for Supernovae
    Rev. Robert Evans 88

3. Photoelectric and CCD Observation — 93

Photoelectric Photometry of Variable Stars
    Douglas S. Hall 95
Robotic Telescopes for Photometry
    John Baruch 109
Multichannel Photometry for Amateur Astronomy Groups
    J.M. Le Contel and E.N. Walker 117
The Joint European Amateur Photometer: An Update
    E.N. Walker 122
Extragalactic Photometry in the Transition Era between Photoelectric
  and CCD Photometry
    G. Longo, G. Busarello and C. Sterken 126

4. Analysis and Interpretation — 135

Variable Stars and Stellar Evolution
    John R. Percy 137
Period Analysis of Variable Stars
    Jan Cuypers 148
Computer Networks in Variable Star Activities
    Veikko Mäkelä and Aarre Kellomäki 156
Highly Variable Objects in the Solar System
    Wieslaw Z. Wisniewski 159

5. Blue and Yellow Variable Stars — 169

The Study of Short-Period Variable Stars through International Cooperation
    Michel Breger 171
Photographic Photometry: Application to Population II Variables
    Martha L. Hazen 185
Visual Detection of the Two Periods of the Double-Mode Cepheid EW Scuti
    A. Figer, E. Poretti and C. Sterken 190
Eta Carinae, AG Carinae, and the Hubble-Sandage Variables
    Roberto Viotti 194
Recent Work on R Coronae Borealis Stars
    David Kilkenny 205
The Hot R CrB Star V348 Sgr and the Evolutionary Status of the R CrB Stars
    Don Pollacco 214

6. Red Variable Stars   219

Observational Perspectives on Red Giants
    F.R. Querci, M. Querci, B. Fontaine and A. Klotz 221
The Analysis of Observations of Mira Stars
    John E. Isles 231
Detecting Period Changes in Mira Variables
    C. Lloyd 242
Predicting the Behaviour of Variable Stars
    Marie-Odile Mennessier 247
International Cooperation for Coordinated Studies of Mira Variables
    Margarita Karovska 255
The Variable Star X Ophiuchi
    Dominique Proust and Michel Verdenet 259

7. Cataclysmic Variable Stars   265

Symbiotic Variable Stars
    Joanna Mikolajewska 267
The Benefit of Amateur Observations for Research in Dwarf Novae
    Constanze la Dous 279
Coordination of Multiwavelength Groundbased and Satellite
    Observations of Novae in Outburst
    Steven N. Shore 290
IUE Observations of Mass-Losing Cataclysmic Variables
    Janet E. Drew 302
TT Crateris 1987-1990
    Richard Fleet 311
Visual Detection of Superhumps in SW Ursae Majoris
    Bjorn H. Granslo 314

8. Closing Remarks   321

9. Index   327

Preface

The first European meeting of the American Association of Variable Star Observers (AAVSO) was held from 24 to 28 July 1990, in Brussels, Belgium. The theme of the meeting - "International Cooperation and Coordination in Variable Star Observing" - reflected the meeting's history, its purpose, and its spirit. For decades, the work of the AAVSO has been international in scope. Each year, over half of the approximately 250,000 observations received by the AAVSO come from outside North America, as do many of the over 200 requests for AAVSO data or services. The AAVSO has made a major contribution to the European Space Agency's HIPPARCOS astrometric satellite project by assisting in the prediction of the brightnesses of several hundred long period variable stars to be observed by the satellite. Thus, a European meeting of the AAVSO was timely as well as pleasant.

The meeting was held at the Vrije Universiteit Brussel (VUB) and hosted by the Vereniging voor Sterrenkunde (VVS), a Belgian association of amateur astronomers which is active in variable star observing. The AAVSO is deeply grateful to Dr. Christiaan Sterken (VUB); Ludwig Cluyse, Johan Gijsenbergs, Christian Steyaert, and Patrick Wils (VVS); and the Local Organizing Committee, chaired by Ludwig Cluyse and including Roland Boninsegna (GEOS) and the individuals mentioned above. They made the meeting truly memorable.

The organization of the meeting, like the light curve of a typical variable star, has its "ups" and "downs". Among the "ups" were the opportunities when the three of us could meet (either in pairs or, occasionally, all together) to work on the planning. Two of us (JAM and JRP) recognize that CS and the Local Organizing Committee had to contend with most the "downs", but they did so very successfully, and it appeared that they ended the meeting in the same "up" state as we did!

The participants numbered over 175 - a nearly equal mix of amateurs and professionals. They came from 30 countries on 5 continents, including a large contingent from eastern Europe. The program consisted of 35 review papers (almost all of them included in these proceedings), and over 50 contributed papers which will appear in a special issue of the Journal of the AAVSO. At the suggestion of the publisher, we have included specially-commissioned papers on the history of variable star astronomy (by JRP) and on variable star observing (by JAM). In addition to the formal discussion after the review papers (also included in these proceedings), informal discussions were an important part of this meeting. Many of them took place in a reception hall containing the poster papers and a splendid exhibit provided by the European Southern Observatory. The participants will also remember the fine Belgium beers which fueled the discussion sessions, the cosmopolitan personality of the city of Brussels, the tours and other social events.

The list of acknowledgements reflects the many facets of the meeting: the Scientific Organizing Committee: M. Breger, C. de Loore, R. Evans, M. Hazen, D. Kilkenny, J.A. Mattei, P. Melchior, J.R. Percy (Chair), D. Proust, C. Sterken,

B. Szeidl, and W. Wamsteker, who planned and contributed to the meeting in various ways; a Regional Organizing Committee consisting of representatives of 22 national variable star observing societies; Professor Adriaan Blaauw, one of Europe's most distinguished astronomers, and former Director-General of the European Southern Observatory, for honouring the meeting with his presence, and for delivering the introductory paper; Professor S. Loccufier, Rector of the VUB, for hosting a concert for participants in Brussels' Spiegelzaal; the VVS members for the friendly opening reception; the director and staff of the Royal Observatory, Belgium, for hosting a tour of the observatory for participants in the meeting. Travel grants for some participants were made possible by donations from : American Astronomical Society, Association of Universities for Research in Astronomy, Astronomy Magazine, Clinton B. Ford, Lichtenknecker Optics, National Fund for Scientific Research Belgium, National Radio Astronomy Observatory, John R. Percy, Sky Publishing Corporation, Universities Space Research Association, and the American Association of Variable Star Observers.

Janet A. Mattei thanks the council, especially President John R. Percy, and Vice-President Martha Hazen, and the staff of the AAVSO, especially Grant Foster, Susan M. Power, Jennifer Rogers, and Elizabeth O. Waagen. John R. Percy thanks Leta Hudson and Mary Magri for their many contributions to the planning and execution of the meeting. The Local Organizing Committee thanks H. Ruland, Vice-President of Lichtenknecker Optics, for his assistance with local arrangements.

The camera-ready copy for these proceedings was prepared at the Erindale Campus, University of Toronto, by Leta Hudson, under the supervision of John R. Percy, who is deeply grateful to Leta, and to Mary Magri, Carmen Cupido and Kim Kubas, for their contributions to this project. We thank Cambridge University Press, especially Simon Mitton, for advice and assistance, and for a grant to cover some of the costs of preparing this lasting record of the Brussels meeting.

What of the future? The International Astronomical Union (IAU), which is responsible for coordinating astronomy around the world, has formally recognized the important role of amateurs in astronomy by passing a special resolution at its 1988 General Assembly, and by establishing a Working Group on Amateur-Professional Cooperation in Astronomy. We hope that the 1990 Brussels meeting of the AAVSO is the first of many such gatherings, and that it - and these proceedings - will help to expand international cooperation and coordination between amateur and professional astronomers, not just in variable star observing, but in all branches of astronomy.

John R. Percy
Janet A. Mattei
Christiaan Sterken

April 12, 1991

Group Photograph

Key to the First European Meeting of the AAVSO Photograph

Participants

1. Tom Sterken
2. Winston Wilkerson
3. Carmen Wilkerson
4. Elena Franciosini
5. Andrea Boattini
6. Robert Evans
7. Antonio Bianchini
8. Jaroslav Kruta
9. John Percy
10. Walter Scott Houston
11. Ronald Royer
12. John Griesé III
13. Charles Scovil
14. Janet Mattei
15. Jaroslava Krutova
16. Wolfgang Quester
17. Bernd-Christoph Kaemper
18. Geert Hoogeveen
19. Miriam Houston
20. Jerzy Speil
21. Steven Padilla
22. Natali Lipunova
23. Hideo Satoh
24. Michel Dumont
25. Jean Gunther
26. Elizabeth Waagen
27. Michel Breger
28. Nikolai Samus
29. Dirk Laurent
30. Sei-ichi Sakuma

List of participants

31. Wim Nobel
32. Melvyn Taylor
33. Peter Serne
34. Igor Kudzej
35. Willy DeKort
36. Weislaw Wisniewski
37. Farouk Mahmoud
38. Axel Thomas
39. Adriaan Blaauw
40. Thomas Cragg
41. Margarita Karovska
42. Francois Querci
43. Cristiaan Steyaert
44. John Baruch
45. Victor van den Bosch
46.
47. Frank Bateson
48. John Isles
49. Liu Zongli
50. Steve Shore
51. Jean-Michel Le Contel
52. Danielle Le Contei
53. Roberto Viotti
54. Harald Marx
55. Michael Collins
56. Stefan Korth
57. Marek Wolf
58.
59. Jiri Borovicka
60. Jindrich Silhan
61. Zdenek Kvis
62. Takashi Iijima
63. Haldun Menali
64. Tamas Zalezsak
65. Clark Neily
66. William Albrecht
67. Behrez Meri-Davan
68. John Pazmino
69. Edward Guinan
70. Myrna Coffino
71. Martha Liller
72. Gerold Richter
73. David Kilkenny
74. Jan Cuypers
75. Howard Landis
76. Danie Overbeek
77. Chris Lloyd
78. Marvin Baldwin
79. Helga Dyck
80. Gerry Dyck
81. Bart Sterken
82. Ed Los
83. Risto Heikkila
84. Aarre Kellomaki
85. Istvan Kovacz
86. Attila Mizser
87. Ludwig Cluyse
88. Christiaan Sterken
89. Malcolm Porter
90. Storm Dunlop
91. Richard Fleet
92. Dietmar Bohme
93. Joanna Mikolajewska
94. Roland Boninsegna
95.
96. Constanze la Dous
97. Patricia Lampens
98. Wayne Lowder
99. Michel Grenon
100. Jean Dommanget
101. Thomas Williams
102. Francisco Pujol
103.
104. John Coggins
105. William Hodgson
106. Louis Cox
107. James Ellerbe
108. Gerry Samolyk
109. Manfred Durkefalden
110. Douglas Hall
111. Robert Reisenweber
112. Johann Gisjenbergs
113. Patrick Wils
114. Koji Mukai
115. Jean-Pierre Sarejan
116. Norman Walker

PROLOGUE:

INTERNATIONAL COOPERATION AND COORDINATION

The European Southern Observatory (ESO) is one of the great examples of international cooperation and coordination in astronomy. One of the highlights of this meeting was a display, from ESO, on astronomy; it was technically and artistically superb. The meeting was also honoured by the presence of Professor Adriaan Blaauw, former Director-General (1970-1975) of ESO, and one of Europe's most distinguished astronomers. He has been a faculty member at major universities in the US and the Netherlands. His research work on OB stars and on star clusters and associations has been recognized by a major symposium on "Birth and Evolution of Massive Stars and Stellar Groups", in honour of his 70th birthday. He has held the highest administrative positions in international astronomy, including President of the International Astronomical Union. He has received many honours, including Knight in the Order of the Netherlands Lion.

The following is the text of the introductory paper which he was invited to give.

INTERNATIONAL COOPERATION AND COORDINATION IN ASTRONOMICAL RESEARCH

Adriaan Blaauw
Kapteyn Laboratorium
Universiteit Groningen
Postbus 800
9700 AV Groningen, Netherlands

Astronomy has a rich history of world-wide collaboration, a history we feel we can be proud of, and about which I feel privileged to speak at the beginning of these meetings. "Cooperation and Coordination" are in the title under which the organizers of the meeting have asked me to speak. Cooperation, "let us do some planning and divide our tasks", so that each of us contributes his building stone for the big project, chosen in accordance with his interest and abilities. The subject is very fitting for this meeting with its participation of both professional and amateur astronomers, for we know that the one cannot do well without the other. What I would like to do, is to have a look with you at a few of the great joint undertakings in the past, but before I do so, may I make a few general and personal remarks.

The first one is that, at an occasion like this, I realize again what a lucky lot we astronomers are - professionals or amateurs - to have chosen this job or this hobby! What made us make the choice was undoubtedly for most of us our curiosity about what is going on among the planets, the sun, the stars and in the immense space beyond. Having then found our way to satisfy this curiosity, and experiencing how over and over again new perspectives are opened and the boundaries of the unknown world recede, is in itself a fantastic reward. But apart from this, there is the other reward we get as an extra bonus, one that, I believe, scientists in few disciplines enjoy as generously as we astronomers do. I might call it the human relations aspect, the fact that doing astronomy brings us in contact with colleagues in many different parts of the world, with different cultural traditions, different languages and sometimes very different political convictions. The meetings we are going to attend this week are one of these unplanned rewards. Speaking for myself when, as a schoolboy, I became so deeply interested in astronomy that I wanted to become an astronomer, I am sure this aspect of the profession never crossed my mind - perhaps it might even have frightened me to become involved in things that world-wide.

Nature forces us to work in world-wide ambience. With so big a universe, containing so many more observable stars and many more observable galaxies than there are astronomers and amateurs, we have for centuries joined forces and often crossed boundaries to get what we wanted. We looked for places with a better climate than our own, we hurried to far-away places to observe a transit of Mercury or Venus, and the Sun and the Moon conspired in sending us here and there to observe the eclipses of the one by the other; perhaps, sometimes, in vain, but then there remained the usually pleasant memory of an interesting trip. A summary in Van de Hulst's chapter on the Chromosphere and Corona in the handbook on The Solar System of 1953, tells us that astronomers swarmed out to places as diverse as Lapland, Peru, Siberia, the Pacific Ocean, Brazil, etc. and amateurs often joined the professionals in this irresistable urge.

I assume that all of you are well aware of the great international projects that nowadays so strongly put their stamp on our work, particularly in the area of space research; international participation in these has become the rule rather than the exception. Let me therefore review a few of the great undertakings in international collaboration and coordination in the past, and show you how strong already in early days was the impetus for the international set up. There were a variety of motives: sometimes it was the necessity to work jointly for assembling in wholesale way certain observational data, sometimes it was a research project, sometimes the necessity to agree on nomenclature or terminology, and sometimes the necessity to pool financial means and manpower to get observational facilities more powerful than what one country could afford.

The "Carte du Ciel" - Astrographic Catalogue

In the second half of the last century, new perspectives were opened by the introduction of photography in astronomy. This led to broad international participation in the project of the Astrographic Catalogue, or Carte du Ciel as it was called in France where the project originated. The brothers-opticiens Paul and Prosper Henry had developed photographic astrographs with excellent definition of stellar images in fields of 2° x 2°, which allowed reaching 14th photographic magnitude with reasonable exposure times. Imagine what this meant for mapping the whole sky! It was a step forward, comparable to the enormous advance now some 40 years ago, when the Palomar Schmidt Survey, and soon after this the ESO/SRC Survey, mapped the whole sky to still fainter limits. Previously, the only overall listings of stars had been the Bonner and the Cordoba Durchmusterungs of which the magnitude limits had been about four magnitudes brighter.

Encouraged by the French government, French astronomers invited observatories all over the world to participate, and after a series of meetings in the years 1887-1891, 18 observatories committed themselves to observe a particular zone in declination. In fact, the project consisted of two parts: one for making maps to be distributed to all observatories and reaching 14th magnitude, the other for making accurate measurements of the stellar positions down to about 12th magnitude and to be published in catalogues, per zone of declination. The two purposes are recognized in the French and the English names.

I wish I could tell you now, that all the observatories that started so enthusiastically have neatly completed their jobs - but that was not the case. The measurement of the positions proved to be such an enormous job that after a while several observatories felt that it tied their hands too much and prevented them from participating in new trends in astronomy. The ones that stuck most conscientiously to their commitment were the French, which is understandable because it had been their initiative. Eventually, these jobs were finished by other institutes. Perhaps, the astronomical world had been too enthusiastic around the year 1900. However, looking back now over almost a century, we note that the "Carte" and the Catalogues have provided us with a document on what the sky looked like around the turn of the century, a document of unique nature. In 1989 the hundredth anniversary of the project was celebrated in Paris with a Symposium "Mapping the Sky; Past Heritage and Future Directions".

The Plan of Selected Areas

The next broad international project I wish to tell something about is the Plan of Selected Areas. Like the Carte du Ciel it aimed at collecting observational data, but in this case these were chosen so as to serve a well defined object of research: the structure of the Milky Way system. Perhaps one should rather say: the structure of the distribution of the stars in space, for at the time the project was drawn up, the true nature of the "spiral nebulae" had not yet been definitely established; we did not know yet that they are stellar systems comparable to our own Galaxy. Of course, a first impression of this structure had been obtained already by William Herschel around the year 1800, on the basis of counts of stars in different directions. A century later, the time was ripe for a new attack, i.e. for collecting the kind of observations rendered accessible by new observational techniques. Photographically, stars of about 19th magnitude had become within reach.

It started in the year 1905 with an initiative by J.C.Kapteyn. He proposed to observe stars in 206 small areas uniformly distributed over the sky so as not to be biased from the outset, and in these the stars' magnitudes, colours, proper motions, radial velocities etc. should be measured to the faintest limits practicable. Altogether an enormous job, that could be done only in collaboration between many observatories. The Plan appealed so strongly to astronomers all over the world, that in the course of time 17 observatories in the northern and southern hemisphere participated. Among these were, to mention a few, the Harvard, Mt. Wilson, Pulkovo, Uppsala and Radcliffe Observatories. The work was first coordinated by Kapteyn himself, but soon became directed by a small committee that became one of the Commissions of the International Astronomical Union in the Transactions of which we find the progress reports over the years.

On the basis of these data, Kapteyn and others determined the structure in the space distribution of the stars. We now know that what they found for the region close to the sun was near to the truth, but that for larger distances it failed, due to the interstellar absorption which one seemed to be unable to adequately measure at that time. The real breakthrough in galactic research occurred around the year 1920, mostly through the work of Harlow Shapley who succeeded in measuring the distances of the globular clusters. Since these appeared to be concentrated around the galactic centre, this centre, far beyond the local region mapped by Kapteyn now was identified. However, the observational data collected in the Plan of Selected Areas remain a lasting document on the properties of the stars, useful even now or in the future.

The International Astronomical Union

Undoubtedly the most comprehensive, and the most influential international undertaking we have in astronomy is the International Astronomical Union. Of course there are also such important international unions as URSI, but the IAU deserves in this context special mention because it is the oldest and embraces all domains of astronomy.

The IAU had its origin here in Brussels, in 1919 just after World War I, when scientists wanted better international relations. They created a number of international unions, but colleagues from certain nations were not immediately accepted due to ill feelings remaining after the war. Fortunately, this was remedied in the course of the decades that followed.

The first General Assembly of the IAU was held in 1922, in Rome, and right away 19 member states joined. The international astronomical community was very small as compared to what it is now: only 207 people constituted the personal membership, but they involved the Union immediately in all aspects of astronomy and formed 32 commissions for the various disciplines. In the course of time, some of these have been abolished, and several others have been added as new fields of research opened up. Of the two concepts in the title of this talk, cooperation and coordination, certainly coordination applies especially to what the IAU does. Its triennial business meetings take care of innumerable matters of planning and coordination without which astronomy might look very messy today. As we all know, the IAU has grown very much since that first Assembly and now has some 6000 members from about 50 member states, who do their work in about 40 commissions.

The IAU, I am happy to say, does even more than matters of coordination in research. It is also concerned about what precedes research: the education in astronomy, something not only necessary in preparation for research, but also as part of the general cultural development of a country. We recognize this in the IAU Commission for teaching, and in the organization of regional meetings which are accessible to many young astronomers who do not yet participate in the more specialized IAU Symposia and in the Assemblies. The IAU-UNESCO International Schools for Young Astronomers are another example of the educational activities. The IAU helps bridge gaps in international relations, and it is perhaps not exaggerated to say, that sometimes it is one step ahead of professional politicians.

Financial Considerations

The above examples have shown how we were motivated scientifically to do things internationally: by the scope of an extensive sky survey made possible by a new technique (The Carte du Ciel), by a common goal in research (The Plan of Selected Areas for Milky Way research), and for reasons of coordination and education (The IAU). Let me conclude with a very down-to-earth reason: our financial means, or rather, the limitation of our finances. It is not something of modern times only, but it does impress us stronger than was the case previously.

As we all know, if we want to reach to the limit of astronomy, it has become a very expensive branch of science. I am not thinking here only of space research - where funding has been strongly motivated by the prospect of developing techniques also applicable in other fields than astronomy - but also of ground based astronomy. Modern techniques render possible the construction of telescopes with light gathering power an order of magnitude larger than what was possible a few decades ago.

Naturally, we feel that, since this now is possible, we must realize such instruments in order to penetrate as deep as we can into the farthest and youngest parts of the universe. And so, we pool financial resources and manpower for the construction of large observatories on an international basis. Examples are the European Southern Observatory in the Chilean Andes, and on the northern hemisphere the Observatory on La Palma, and they are very successful. It would seem that, apart perhaps from the United States, no individual country can afford any more the establishment of an observatory that makes full use of what modern technique offers. We are forced to collaborate in international teams for both the design of instrumentation and for drawing up the new observing projects, like, for instance, ESO's Key Projects. And I am sure that here, again, we welcome and enjoy this, now "forced", internationalization.

CHAPTER 1

HISTORY AND ORGANIZATION

VARIABLE STARS: A HISTORICAL PERSPECTIVE

John R. Percy
Erindale Campus
University of Toronto
Mississauga, Ontario
Canada L5L 1C6

Sometime, far back in prehistory, people first began watching the sky. They must certainly have noted the daily and yearly motions of the sun, the waxing and waning of the moon, and the wanderings of the planets. They may even have noted changes in the stars: not changes of *position* but changes of *brightness*. There are hundreds of **variable stars** among the naked-eye stars, and the changes of some of them are so striking that they could not escape the notice of a careful skywatcher.

Written records of skywatching begin in the Near East around 2000 B.C., and in the Far East around 1000 B.C. Unwritten records also exist in many forms: paintings, rock and bone carvings, alignments of giant stones. The study of these has been called "archaeoastronomy", and this has developed into an interesting, popular and somewhat controversial subject. The interpretation of any ancient records is a challenge to both the astronomer and the historian, a challenge which is best met by truly interdisciplinary study.

The Babylonians laid the foundations of Western astronomy through their mathematics and through their systematic observations of the sun, moon, planets and stars. We can forgive the fact that they did this for astrological purposes! Babylonian astronomy was later absorbed into Greek culture, where it eventually became part of "models" of the visible universe.

In some schools of philosophy, these models were intended as mathematical conveniences, designed only to represent or predict solar, lunar or planetary motions. In other schools, these models took on wider significance: they were intended to represent physical reality. Aristotle (384 - 322 B.C.), for instance, wrote that the world was made of four elements: earth, water, air and fire. Bodies beyond the sphere of the moon are made of a fifth element or "quintessence" which was ingenerable and incorruptible, and underwent only one kind of change: uniform motion in a circle.

Aristotle's works had a tremendous impact on Western thought. They survived in translation even after the original versions (such as those in the famous library at Alexandria) were lost. Since Aristotle said that stars can't vary, there are few if any records of variable stars in Western literature. An exception is the observation of a nova by Hipparchus in 134 B.C. The "chronicles" maintained by monasteries in mediaeval times record many astronomical and meteorological phenomena, but variable stars are not among them. There is only one European record of the "new star" of 1006 A.D. (now classified as a **supernova**), the brightest such event on record.

In the Orient, observers were not inhibited by Aristotle. In fact, their cultures often placed great importance on omens, such as unexpected events in the sky. Chinese, Japanese and Korean records are a fruitful source of information on supernovae, novae, comets, eclipses and other such events, if one can interpret the records correctly. It appears that about 80 "new stars" were recorded up to about 1600, 8 being supernovae and the rest being ordinary novae. Other cultures, in the Middle East and in North America, may also have recorded these "new stars", but they did not do so in the precise, methodical fashion of the Orientals.

Tycho's and Kepler's Stars

In the West, the sixteenth century brought a renaissance and revolution in scientific thought. (In fact, the word *revolution*, used in the sense of the overthrow of world order, comes from the word *revolution* in the title of Copernicus' famous book, used in the sense of motion about the sun!) Ironically, the seeds were contained in the writings of Aristotle: By contributing to the basis of the scientific method, he developed the weapons with which he was later attacked and overthrown.

By happy coincidence, as the revolution of ideas was taking place, Nature provided two rare events -- supernovae -- which would contribute to the demise of Aristotelianism. In November of 1572, Tycho Brahe (1546 - 1601) recorded a "new star" in Cassiopeia. He (and independently the English astronomer Thomas Digges) established its position and its fixed, stellar nature. They also measured its changing brightness relative to the planets and stars. At maximum brightness, it rivalled Venus! He published his results in a book entitled "On a New Star, Not Previously Seen Within the Memory of Any Age Since the Beginning of the World". Tycho Brahe's observations were so careful and systematic that later astronomers (notably Walter Baade) could reconstruct the changing magnitude of "Tycho's star" to an accuracy of $\pm\ 0^m2$. Since the distance to the remnant of the supernovae can now be determined, the *absolute* magnitude of the supernova at maximum brightness can be found. This provides a bench mark which has been used to determine the distance to other supernovae.

Barely 30 years later, in October, 1604, Johannes Kepler (1571 - 1630) recorded another "new star". He (and independently David Fabricius, pastor in Osteel in Ostfriesland) established its position and its fixed, stellar nature, and recorded its changing brightness. Kepler's star was somewhat fainter than Tycho's, but still at maximum brightness it rivalled Jupiter.

Only a few years later, the telescope was first applied to astronomy by

Galileo Galilei (1564 - 1642), and since then, astronomers' technical resources have increased by leaps and bounds. Unfortunately, no supernova has been seen in our galaxy since 1604, though a bright (naked-eye) supernova in the Large Magellanic Cloud was discovered in 1987 by Ian Shelton.

The Beginnings of Modern Astronomy

During the next two centuries, variable stars continued to be discovered and observed sporadically, occasionally through deliberate, systematic measures but more often by chance. In August of 1596, David Fabricius discovered a third-magnitude star which he could not find in any star catalogue, atlas or globe. A few months later, it had faded. In 1603, Bayer -- unaware of Fabricius' discovery -- listed this same star in his Atlas as o Ceti. It was not until 1638, when this star was found to be visible at some times but not at others, that it was recognized to be "permanently" variable. By 1660, its periodicity of 11 months was established. The star became known as *Mira*, the wonderful. It is the prototype of one of the most important classes of variable star. Two other members of this class were discovered soon after: χ Cygni in 1686 and R Hydrae in 1704.

Towards the end of the eighteenth century, another burst of progress was made. Some of this must be attributed to William Herschel (1738 - 1822), the "father" of modern stellar astronomy. His development of large reflecting telescopes, his efforts to define, measure and catalogue the brightnesses of stars, and his interest in the physical nature of stars and nebulae, all contributed to the development of the study of variable stars. It is interesting to read what his son John Herschel (1792 - 1871), also a noted astronomer, said about variable stars in the 1833 edition of his "Principles of Astronomy":

> "... this is a branch of practical astronomy which has been too little followed up, and it is precisely that in which amateurs of the science, provided with only good eyes, or moderate instruments, might employ their time to excellent advantage. It holds out a sure promise of rich discovery, and is one in which astronomers in established observatories are almost of necessity precluded from taking a part by the nature of the observations required. Catalogues of the comparative brightness of the stars in each constellation have been constructed by Sir William Herschel, with the express object of facilitating these researches ..."

These words are still true today. At the time that they were written, of course, many eminent scientists were amateurs in the sense that they were not paid to do astronomy. William Herschel was a good example. So too were John Goodricke (1764 - 1786) and Edward Pigott (1753(?) - 1825), two well-to-do Yorkshire aristocrats who made a special study of variable stars. Pigott is credited with the discovery of the variability of η Aql in 1784, of R CrB in 1795, and of R

Sct in 1796. Goodricke is credited with the discovery of the variability of δ Cep and β Lyr in 1784, and of the periodicity of β Per (Algol) in 1782 - 1783, (though the periodicity was apparently discovered independently by a German farmer named Palitzch, who had an interest in astronomy). Together, Goodricke and Pigott also proposed that the variability of Algol might be caused by eclipses (but by a planet!). Goodricke, a deaf mute, died at the untimely age of 21, and Pigott, after the death of his friend, gradually gave up astronomy, but during the brief period of their collaboration, "the Yorkshire astronomers" laid the foundations for the study of variable stars as a branch of astronomy.

Systematic Visual Observations

In 1786, Pigott has listed 12 definite variables and 38 suspected ones. By about 1850, F.W.A. Argelander (1799 - 1875) listed 18 definite variables and numerous suspected ones. Furthermore, many of the definite variables had good light curves, thanks to the techniques of measurement which William Herschel had devised and which Argelander had refined. Argelander's work marks another high point in the study of variable stars. He prepared the *Uranometria Nova* (1843), a set of charts which provided improved magnitudes for naked-eye stars, and later his monumental *Bonner Durchmusterung* catalogue (1862) and charts (1863), containing 324,189 stars down to ninth magnitude. In so doing, he made hundreds of thousands of visual measurements of the brightnesses of stars, providing magnitudes for comparison stars and discovering many new suspected variables in the process.

Not the least of Argelander's contributions was his well-known appeal to "friends of astronomy". It appears at the end of an article in Schumacher's *Astronomisches Jahrbuch* for 1844, in which Argelander discusses the history, importance, methods of study, and idiosyncracies of variable stars:

> "Could we be aided in this matter by the cooperation of a goodly number of amateurs, we would perhaps in a few years be able to discover laws in these apparent irregularities, and then in a short time accomplish more than in all the 60 years which have passed since their discovery.
>
> Therefore do I lay these hitherto sorely neglected variables most pressingly on the heart of all lovers of the starry heavens. May you become so grateful for the pleasure which has so often rewarded your looking upward, which has constantly been offered you anew, that you will contribute your little mite towards the more exact knowledge of these stars! May you increase your enjoyment of combining the useful and the pleasant, while you perform an important part towards the increase of human knowledge, and help to investigate the eternal laws which announce in endless distance the almightly power and

wisdom of the Creator! Let no one, who feels the desire and the strength to reach this goal, be deterred by the words of this paper. The observations may seem long and difficult on paper, but are in execution very simple, and may be so modified by each one's individuality as to become his own, and will become so bound up with his own experiences that, unconsciously as it were, they will soon be as essentials. As elsewhere, so the old saying holds here, "Well begun is half done," and I am thoroughly convinced that whoever carries on these observations for a few weeks, will find so much interest therein that he will never cease. I have one request, which is this, that the observations shall be made known each year. Observations buried in a desk are no observations. Should they be entrusted to me for reduction, or even for publication, I will undertake it with joy and thanks, and will also answer all questions with care and with the greatest pleasure."

[This translation, by Annie J. Cannon, appears in *Popular Astronomy* for 1912, and in "A Source Book in Astronomy".]

Argelander's appeal did not go unheeded. The Variable Star Section of the British Astronomical Association was founded in 1890 and has been active ever since. In America, the study of variable stars was supported especially by E.C. Pickering (1846 - 1919) at the Harvard College Observatory. In 1882, Pickering had published "A Plan for Securing Observations of Variable Stars", pointing out the scientific value of visual observation of variable stars. In 1911, the American Association of Variable Star Observers was founded in Cambridge, Massachusetts. Originally, it was associated with the Harvard College Observatory. Since 1954, it has been a completely independent organization. It is a truly international one, receiving hundreds of thousands of observations each year from members and associates around the world.

Pickering had pointed out in 1882 that the study of variable stars might especially appeal to women. This has proven to be the case, in part because Pickering and later Harlow Shapley encouraged and hired women astronomers. Annie Cannon, Dorrit Hoffleit, Henrietta Leavitt, Margaret Mayall, Cecilia Payne-Gaposhkin and Helen Sawyer Hogg are among many notable examples of women who have made important contributions to our understanding of variable stars.

The Photographic Revolution

The development of photography in the late nineteenth century had a profound effect on the study of variable stars. This had begun many years earlier (John Herschel being one of the pioneers), and was first applied to stellar astronomy in 1850, but it was not until about 1880 that emulsions became sensitive enough, and telescope drive systems became accurate enough, for astrophotography to make its full impact. The photographic plate, which could now integrate starlight for up to several hours, could record stars which were far too faint to be seen visually in the telescope. Furthermore, thousands of stars could be recorded on a single plate, making discovery and measurement more efficient. Photography provided a permanent record of star brightness, capable of being checked or remeasured at any time. If a newly-discovered variable had been recorded on earlier plates, its previous history could be investigated. Archival collections such as the plate file of the Harvard College Observatory contain a gold mine of information on the past variability of stars -- and even of more exotic objects such as quasars.

Many observatories embarked on systematic photographic surveys of variable stars, notably Harvard, Heidelberg, Leiden, Sonneberg and Sternberg. Each observatory accumulated its own collection of variables, with consequent problems of nomenclature. Selected regions of the sky were more closely surveyed: the northern and southern Milky Way, the region of the galactic nucleus in Sagittarius, globular clusters, the Magellanic Clouds, and later more distant galaxies such as M31. The number of known variables increased rapidly to over 1000 in 1903, over 2000 in 1907, and over 4000 in 1920. Today, there are well over 30,000 confirmed variable stars.

Spectroscopy

The development of astronomical spectroscopy and spectral classification in the late 1800's led to a better understanding of variable stars. Unusual aspects of the spectra of variable stars were noted visually as early as 1850: the absorption bands in the spectra of cool, long-period variables, and the bright emission lines in the spectra of novae and of Be stars such as γ Cas. The development of photographic spectroscopy by H. Draper in 1872 led to large-scale projects in spectral classification, notably by Pickering and his associates at the Harvard College Observatory. It became possible to classify variables according to temperature, and later according to luminosity; this enabled astronomers to learn how variable stars compared and fit in with normal stars. This in turn led to improvements in the classification schemes for variable stars.

Classification and Explanation

The earliest classification scheme dates back nearly two centuries: Pigott divided variables according to the nature of their light curve into novae, long-period variables and short-period variables. A century later, Pickering devised a more detailed scheme: (Ia) normal novae, primarily nearby ones in our own galaxy (Ib) novae in nebulae: now known to be primarily *super*novae in distant galaxies (IIa) long-period variables: cool, large-amplitude pulsating variables (IIb) U Geminorum or SS Cygni stars: dwarf novae (IIc) R Coronae Borealis stars: stars which suddenly and unpredictably decline in brightness (III) irregular variables: a motley collection (IVa) short-period variables such as Cepheids and later including the cluster-type or RR Lyrae stars (IVb) β Lyrae type eclipsing variables and (V) Algol type eclipsing variables.

Pickering's classification scheme contains some hints as to the nature and cause of the variability. Classification and explanation go hand in hand. In the late nineteenth and early twentieth century, progress was made in understanding the physical nature and the physical processes in variable stars, and in stars in general. This culminated in Arthur S. Eddington's (1882 - 1944) monumental book "The Internal Constitution of the Stars".

Two centuries ago, Goodricke and Pigott had speculated that the variability of Algol and similar stars might be due to eclipses. A century later, the eclipse hypothesis was firmly established: in 1880, Pickering carried out a mathematical analysis of the orbit based on careful observations of the light curve, and by 1889 the orbital motion was directly observed by H.C. Vogel using astronomical spectroscopy and the Doppler effect. Nevertheless, some eclipsing variables defied explanation and, even today, Goodricke's β Lyr is one of the most enigmatic objects in the sky. Furthermore, the eclipse hypothesis was often overextended: it was used to explain the Cepheids, long-period variables and even the R CrB stars! When the pulsating variables were finally understood, the pendulum swung the other way, and the importance of binarity in explaining the variability of such other types as novae and dwarf novae has only recently been appreciated.

In a sense, Goodricke anticipated yet another type of variability. Unable to explain the behaviour of β Lyr under the eclipse hypothesis, he suggested that "the phaenomenon seemed to be occasioned by a rotation on the star's axis, under a supposition that there are several large dark spots upon its body, and that its axis is inclined to the earth's orbit". Although this explanation is not directly relevant to β Lyr, it does apply to rotating variables such as the magnetic peculiar A stars, and the RS Canum Venaticorum stars.

The idea that variability might be due to pulsation was raised (by A. Ritter) as early as 1873, but it was not until the observational studies by Harlow Shapley (1855 - 1972) and others around 1915, and the concurrent theoretical studies by Eddington that the pulsational nature of the Cepheids, cluster type variables and the long-period variables was established. The cause of the pulsation -- the thermodynamic effects of hydrogen and helium in the outer layers of the stars -- was not firmly established until after 1950.

Photoelectric Photometry: The Electronic Revolution

Around the beginning of the twentieth century, another important technique was added to the arsenal of the variable star astronomer: photoelectric photometry. The technique apparently originated in Great Britain. At first, it was insensitive and cumbersome and could only be applied to the planets and the brightest stars. However, through the careful efforts of its first practitioners, J. Stebbins (1878 - 1966) and P. Guthnick (1879 - 1947), this technique was successfully used to discover the shallow secondary eclipse of Algol (1910), the small light variations in the pulsating β Cephei stars (1913) and in the rotating magnetic stars such as α^2 CVn (1914). This technique has since revealed several other new classes of "microvariables". It has also enabled small colour variations to be observed. There is a lesson to be learned here: Stebbins and Guthnick were pioneers; they knew the limitations of their technique, and they took great pains to apply it carefully. Nowadays, photoelectric photometry is routine, and its practitioners often apply considerably less care than Stebbins and Guthnick did. The result is often mediocre, inferior data. On the other hand, photoelectric photometers are now simple and inexpensive enough for amateur use. With care, the amateur can now make important contributions to the photoelectric photometry of variable stars, particularly those with unpredictable, irregular or slow variations.

A second "electronic revolution" has occurred in the last two decades of the 20th century. Electronic detectors (notably charge-coupled devices or CCD's) have been miniaturized and are now available in arrays of tens of thousands of pixels. Electronic imaging has many advantages over conventional photography, notably much greater sensitivity and linearity of response. Within the last year or two, the prices of CCD's have decreased substantially, and they are now within the reach of groups of amateur astronomers, and well-to-do individuals.

During this century, the basic tools of variable star astronomy - visual, photographic and photoelectric photometry, spectroscopy and physical analysis - have gradually been refined, and have produced a steady stream of important results. Many of these are alluded to in this book.

The Modern Age

In the last few decades, several sophisticated new techniques have opened up new windows on the universe, allowing us to understand variable stars -- and all other celestial objects -- more deeply. One technique is the use of radio telescopes. The discovery of pulsars was probably the most unique and important contribution of radio astronomy to any field of stellar astronomy. Radio telescopes can also probe the interstellar matter around forming stars, as well as the matter ejected from supernovae and other eruptive stars.

Infrared (IR) astronomy began in the 1920's using thermocouple detectors. Since then, detectors and telescopes have improved to the point where we can now observe IR radiation from stars such as Cepheids, even in external galaxies. IR telescopes are particularly useful for studying cool variables (which emit most of their energy in the IR) as well as for probing the cool gas and dust around stars, both old and young. IR detectors have become simple and cheap enough to be within the reach of the amateur, and there is particular need for systematic, long-term IR studies of variable stars -- studies which can best be carried out at smaller, local observatories.

Ultraviolet (UV) astronomy must be carried out from above the atmosphere, because the ozone layer absorbs all wavelengths shorter than 3000 Å. The first rocket observations were made in 1955, but UV astronomy truly blossomed with the launch of UV satellites: Orbiting Astronomical Observatory 2 in 1968, Orbiting Astronomical Observatory 3 ("Copernicus") and the European Space Research Organization's TD1 in 1972 and the International Ultraviolet Explorer (IUE) in 1978. The IUE has been the "workhorse" of UV astronomy, producing thousands of astronomical research papers in the first decade of its life. These UV satellites have been especially effective in studying hot stars (such as the Be stars) and hot gas (such as the chromospheres of "active" stars and the discs of gas around the dense components of close binary systems).

X-ray astronomy must likewise be carried out from above the atmosphere, and has likewise been revolutionized by the launch of satellites, particularly Uhuru in 1970 and the High Energy Astrophysical Observatory (HEAO) A in 1977 ad HEAO B ("Einstein") in 1978. Perhaps the most significant discoveries of X-ray astronomy have been the X-ray emission from close binary systems with one normal component and one collapsed component: a white dwarf, neutron star or black hole. In these systems, gas flowing from the normal component into the gravitational field of the collapsed component becomes heated to millions of K. Cataclysmic variables, which are systems of this kind, have been particularly well studied by this technique.

Finally, we should not forget another powerful tool of modern astronomy: the electronic computer. It allows us to deal with the millions of "bits" of data sent back from space observatories, or generated by modern electronic detectors on ground-based telescopes. It also allows us to model the complex physical processes involved in stellar variability: pulsation and explosion, mass transfer in binary systems, and flares on stellar "surfaces". Some of these problems require the world's most powerful computers. At the other end of the scale, inexpensive microcomputers are available to automate virtually every aspect of the data acquisition and reduction process, in amateur as well as professional astronomical observatories.

Variable Stars: The Present Status

The study of variable stars is one of the most popular and dynamic areas of modern astronomical research. Variability is a property of most stars, and as such it has a great deal to contribute to our understanding of them. It provides us with additional parameters (time scales, amplitudes ...) which are not available for non-variable stars. These parameters can be used to deduce physical parameters of the stars (mass, radius, luminosity, rotation ...) or to compare with theoretical models. We can choose a "model" of a given star by requiring that it reproduce the observed variability as well as the other observed characteristics of the star. From the model, we can then learn about the internal composition, structure and physical processes in the star.

The study of variability also allows us to observe changes in the stars directly: both the rapid and sometimes violent changes associated especially with stellar birth and death, and also the slow changes associated with normal stellar evolution.

It is tempting to think that we have now discovered and studied all types of stellar variability, but history constantly proves us wrong. Even within the last few years, new types of variability have been discovered, usually (but not always) as a result of the development of new astronomical tools and techniques. Until astronomers have studied all types of stars, on all time scales from milliseconds to centuries, in all regions of the spectrum, the search will not be complete. Even then, it is unlikely that we shall understand all that we see.

As a branch of astronomical research, the study of variable stars has a unique breadth. While some astronomers, using the most sophisticated instruments of space and ground-based astronomy, continue to make important discoveries at the frontier of the science, other astronomers (professional and amateur), using much simpler techniques, make equally-important discoveries. It is this breadth which makes the first European meeting of the AAVSO so scientifically exciting for all of us.

SOME LITTLE KNOWN VARIABLE STAR ASTRONOMERS FROM AROUND THE WORLD

Thomas R. Williams
1750 Albans Road
Houston, TX 77005-1704,
USA

Introduction

Amateur astronomy is rich with opportunities for individuals with limited resources to make substantive scientific contributions. Noteworthy contributions have been made over the years, and around the world, by thousands of individuals who have received little notice for their efforts. This paper is intended to highlight a few outstanding amateur contributors to variable star astronomy.

Two types of "little-known" variable star astronomers will be considered, those who are not well known outside their own country or organization, and those who may be well known for contributions to astronomy in fields other than variable stars. The AAVSO has always been an international organization; its charter membership was "from around the world". When Dave Rosebrugh summarized achievements of AAVSO observers in 1946, fully one third of those observers who had contributed more than ten thousand observations each were from countries outside North America. [1]

Amateurs - Who Qualifies?

This discussion will focus on those variable star astronomers who are properly classified as amateurs. This distinction has been discussed in other forums[2] and an extended discussion here is inappropriate. The key characteristics are that an individual be appropriately considered an astronomer who contributed to astronomy after the year 1800 and made substantive contributions while not in a paid position. Also, astronomers who are still living will not be discussed, although there are literally hundreds of amateur astronomers living today, many of them in this meeting, who might very well deserve recognition in this context.

It is convenient to divide variable star contributions into three categories. First and perhaps foremost, there is the work of observation. A second area in which amateur astronomers have made substantial contributions is in the discovery of variable stars and novae. And finally, those few individuals who have contributed substantively to the organization and analysis of variable star observations and to theoretical developments are considered as a third category.

Observation - The Basic Contribution

The dominant contribution of amateur astronomers to variable star astronomy has always been, and is likely to continue to be, in the routine observation of variable stars. The most noteworthy observers are summarized in Table 1 (see end of text). That variable star observation is an international affair is clearly demonstrated with 23 countries around the world providing at least one prominent observer for Table 1.

Two of the three leading AAVSO observers, Reginald de Kock and Cyrus Fernald, were not well known outside their native countries. Yet each made very substantive contributions to astronomy in addition to their outstanding variable star observing efforts. The third AAVSOer of note in that regard, Leslie Peltier, is somewhat better known because of his discoveries of comets and novae in addition to his work as an observer of variable stars. All have been discussed elsewhere and further discussion will not be included in this paper. [3]

C. F. Butterworth

The career of the English variable star observer Charles Frederick Butterworth (1870-1946) deserves highlighting. Butterworth was a dedicated and very disciplined observer in the early morning hours. This added greatly to the value of the nearly 80,000 observations he contributed to the British Astronomical Association Variable Star Section (BAA-VSS), and to the Association Francaise d'Observateurs d'Etoiles Variables (AFOEV).

Butterworth was also a member of the Spectroscopic Section of the BAA and contributed several papers to its journal based on work he did with a 6½ inch prismatic camera. For his work on variable stars Butterworth was honored several times. He received the first Abbot Silver medal from the University of Lyon, the President of France's Palmes d'Officer de l'Académie and the BAA Walter Goodacre Medal and Prize. [4]

G. E. Ensor/H. E. Houghton

Observations from the southern hemisphere are always welcome in the AAVSO archives. In addition to Reginald de Kock, who was already mentioned, there are two other important contributors from South Africa.

George Edmund Ensor (1873-1943) was Director of the Astronomical Society of South Africa Variable Star Section (ASSA-VSS) until 1934. A New Zealander by birth, Ensor was employed as a radiographer at the Pretoria General Hospital. Best remembered for his independent discoveries of two comets Ensor was also active as a lunar occultation observer. Over the period from 1926 to 1940 Ensor contributed 14,952 observations of southern variable stars to the AAVSO archives. [5]

Hendon Egerton Houghton (1892-1947) who succeeded Ensor as Director of the ASSA-VSS, was an official in the High Commissioner's office in Capetown. Houghton migrated from England as an adult in 1920, and was Director of

ASSA-VSS from 1934 until his death. Houghton's variable star observing extended from 1926 to 1942 over which period he reported a total of 23,589 observations to the AAVSO. Houghton was co-discoverer (with Ensor) of Comet 1932 I. He also served as ASSA President, ASSA Secretary from 1923 to 1930, and was active as a contributor of historical articles to the ASSA Monthly Notices. [6]

R. G. Chandra

Among the more dedicated observers in AAVSO history is Radha Govinda Chandra (1881-1975), who observed for many years under adverse conditions in Jessone, Bagchar, India. Chandra became interested in astronomy at the age of 15 when he read a book titled "How Great Is The Universe". He was self-tutored from several astronomy books (in Sanskrit and Bengalee) over the next few years, in the process acquiring binoculars and observing Comet Halley. With encouragement from his father and uncle he purchased a three inch refractor to continue his studies, gradually becoming interested in double, colored and variable stars, star clusters and nebulae. However, when Nova Aquilae "suddenly blazed out in our sky at 9 P.M. Jessone mean time on 7th June 1918....I was anxious to know the particulars of the Nova and I wrote to late Proff. E. C. Pickering". After further correspondence with Pickering and Solon I. Bailey, Chandra was invited to join AAVSO. [7]

Chandra completed his education at the Jessone Government school, and became "Treasurer, Jessone Collectorate" for the Indian Government, while continuing his active interest in variable stars. In 1926 when Charles Elmer donated his six-inch refractor to the AAVSO, this instrument was promptly loaned to Chandra to increase the range of his observing. In 1947 Chandra received an Honorary Membership in AAVSO recognizing his many years of dedicated observing. His last observations were submitted to AAVSO in 1954, bringing his life-time total to 37,215 observations. The number of observations is impressive, but even more important was his continuous record of observations at a longitude far from that of most observers, greatly increasing the temporal completeness of the observational records for the stars he observed. Chandra was also a member of, and contributed observations to, the BAA-VSS and the AFOEV.

G. B. Lacchini

An early 1911 Popular Astronomy article by William Tyler Olcott described opportunities for amateur astronomers to make a valuable contribution to astronomy. Giovanni Battista Lacchini (1884-1967) responded immediately from Italy. He wrote to Olcott on October 12, 1911 and became one of the Charter Members of the AAVSO. By the time AAVSO was incorporated formally in 1918, Lacchini had acquired 350 charts and had reported about 10,000 observations to Olcott, an auspicious beginning to a long and distinguished career in variable star astronomy.

Lacchini was born in Faenza, Italy where he received all his education. He was appointed book-keeper for the Mixed Court in Cairo, Egypt but two years later was recalled to Italy as a postal worker. His initial interest in astronomy was stimulated by reading various works by Camille Flammarion. As an observer, Lacchini's interests were diverse and included comets and minor planets in addition to variable stars, but it was his early observations of variable stars that brought him international recognition. In 1922 Lacchini was elected to the Variable Star Commission of the International Astronomical Union, partly in recognition of his continuous and invaluable record of observations of Z Camelopardalis and T Orionis. Lacchini's contributions eventually led to his appointment to the staff of various Italian observatories, first at Catania, then to Turin University (Pino Torinese Observatory) and finally to Trieste.

In 1952 Lacchini retired from his post as astronomer at Trieste and returned to Faenza. He remained active in astronomy, supporting the Faenza School Observatory which he had helped found in 1921, and working with Guido Horn d'Arturo, Director of the Bologna and Lojano Observatories. During the twilight years of his career, Lacchini discovered numerous variable stars photographically, analyzing plates taken by Horn d'Arturo and himself on the 1.8 meter telescope.

Lacchini was an "active" member of AAVSO in spite of his great distance from the Headquarters in Cambridge. In 1930 he was elected to serve a two year term as a member of the AAVSO Council. At the Fall Meeting that same year, the members of the Council further honored Lacchini by electing him to the position of 2nd Vice President. These recognitions were a reflection not only of his efforts as an observer, but also his assistance in sky checking AAVSO charts for errors. Lacchini contributed 58,812 observations to AAVSO, and in 1952 was elected an Honorary Member of AAVSO. [8]

We have already mentioned quite a few of the leading variable star observers from around the world, but there are two observers from the United States who should be discussed in this context. Both were afflicted with medical problems and died at a relatively young age. In spite of the health complications involved, each managed to make outstanding observational contributions. In addition, both were able to inspire other observers through their enthusiasm, dedication and communication skills.

Curtis E. Anderson

By profession an engineer, Curtis E. Anderson (1927-1977) was afflicted with Multiple Sclerosis (MS). At the age of only 34 Anderson was confined to a wheel chair for the remainder of his life. In spite of his confinement and years of intense pain, Anderson compiled an outstanding record as a variable star observer. He never missed submitting monthly reports to the AAVSO over the twenty-five years of his membership, amassing a total of 60,612 observations. In his later years Anderson was frequently hospitalized for extended periods. However, his indomitable spirit forced the hospital staff to lift him to a window with binoculars so he could make at least one observation each month. Anderson was also well known as editor of the Inner Sanctum column in the informal newsletter Variable Views (VV's). [9]

Carolyn J. Hurless

As the most active woman observer in AAVSO history, Carolyn J. Hurless (1934-1987) would merit inclusion among leading variable star astronomers. However, her value to variable star astronomy goes well beyond the 78,876 observations she submitted to the AAVSO. Hurless was a source of inspiration for literally hundreds of AAVSO'ers over the years. With infectious enthusiasm, outstanding teaching skills and an engagingly warm personality she was able to excite and encourage fledgling observers of all ages.

Hurless carried on a substantial correspondence with observers from all over the world. The numerous letters Hurless received formed the basis for an informal newsletter <u>Variable Views</u>. On a regular basis, mainly monthly, Hurless would gather up interesting excerpts from letters from all her correspondents, and patch these together with cartoons about astronomy, funny quotes about almost anything that struck her fancy, and occasional articles from other publications. These formed the raw material from which she produced a 10-20 page newsletter on a spirit duplicator. <u>Variable Views</u> was mailed to as many as 200 "subscribers" who helped defray some but not all of the cost. <u>Variable Views</u> was a labor of Hurless' love for astronomy and variable star observers. VV's was an effective communications tool that kept an inner circle of variable star observers in touch with others of similar interests and provided constant encouragement and inspiration to all. Hurless served several terms on the AAVSO Council and was an active participant in variable star astronomy in many ways. Thus, her death at the relatively young age of 53 years left a substantial void. [10]

Discovery of Variable Stars

The prospect of discovering a new star is likely one of the motivations that lurks in the backs of the minds of many regular variable star observers. Most variable star observers have discovered at least one "nova" that turned out to be a relatively bright star or variable not yet plotted on a chart. Such discoveries may be either accidental or part of a planned program; history has given us ample lists of both types of discovery events. Table 2 (see end of text) lists a few of the amateurs who have discovered variable stars, novae or supernovae.

An example of an accidental discovery is one of the earliest observations of the extra-galactic supernova S Andromedæ. An independent discovery of this object was made during a casual examination of the galaxy M 31 by the Baroness Berta Degenfield-Schomberg (1884-1928) at her family observatory at Kiskartal, Hungary[11]. These observations were not publicly announced until well after the first formal announcements of the supernova. Identification of the new star by Degenfield-Schomberg was confounded by moonlight. Further, no attempt was made to quantify either the brightness or location of the "new" star. Thus, this discovery has never been fully acknowledged as the "first" observation of S Andromedæ.

Rather than focus on the accidental discoveries, however, and there are many more examples to discuss, I would prefer to concentrate on the results of conscious effort to discover variable stars as an explicit part of an observing program.

T. D. Anderson

Discoveries that could be characterized as less accidental were made by Rev. Dr. Thomas David Anderson (1853-1932) of Edinburgh. Anderson was so familiar with the constellations that he could spot any discrepancy in stars of third magnitude or brighter. Through this ability he discovered Nova Aurigæ (1892) and Nova Persei (1901). In the case of Nova Persei, Anderson's discovery and timely reporting allowed astronomers, for the first time, to observe a nova while it was still brightening. The nova was undetectable no more than 28 hours earlier on photographs taken by Arthur Stanley Williams.

The experience of discovering Nova Aurigæ stimulated Anderson to further efforts at discovery. He undertook a very systematic program to that end, using binoculars, a 2¼ inch refractor and the Bonn Charts. His efforts were rewarded in 1893 with the discoveries of T Andromedæ and V Cassiopeiæ. Anderson was credited with the discovery of at least 46 variable stars. [12]

T. H. Astbury

Another excellent example of a conscious effort to discover variable stars came about as a result of a program to search for novae. In 1903 the BAA-VSS established a nova search program and Thomas Hinsley Astbury (1858-1922), was assigned a portion of the sky to keep watch over. Astbury's systematic and careful approach to this problem with somewhat limited resources was soon rewarded with the discovery of RT Aurigæ, a cepheid variable with a period of only 3.75 days and limited range (5.1 to 5.8 mag). A visual discovery of a variable star with these characteristics can only be characterized as a remarkable accomplishment. Astbury went on to record the discovery of six other variable stars through his efforts including RS, Z, and RT Vulpeculæ, VW Draconis, TV Cassiopiae and W Ursae Majoris. [13]

A. Brun

Many comparison charts were prepared for AFOEV observers by the founder of that organization, Antoine Brun (1881-1978). Brun also published what was likely the first photometric star atlas for use by variable star observers. Brun made many discoveries as a result of his routine observing program starting with his discovery of the variability of SZ Cephei in 1913. Brun observed variable stars regularly while on duty, even in the trenches, with the French Army in World War I. This dedication contributed to an unbroken string of Z Camelopardalis observations over the period from 1914 to 1923 that led to his discovery of the peculiar nature of its variability. Z Cam had been classified as a Mira-type long period variable before his publication of observations in 1923. Brun also was responsible for discovery of a number of variable stars photographically using plates taken with the Schmidt camera at Haute Provence Observatory in France, including 13 discoveries reported at the age of eighty-nine! [14]

A. S. Williams

The concept of exploiting photography as a tool to discover variable stars first occurred to Arthur Stanley Williams (1861-1938) while on a trip to Australia during the winter of 1885-86. It was during this trip that he made his first visual discovery of a variable star, the eclipsing binary star V Puppis. Williams' contributions to planetary science are well known, and his commitment to that work delayed the initiation of a formal photographic program for variable star discovery.

In 1901 Williams began a campaign of photographic examination of various zones of the sky. He used a 4.4-inch portrait lens attached to his equatorial mounting which he drove by hand. Comparison of the exposed negatives from several nights was sufficient to enable Williams to discover over 50 variable stars including Y Lyræ, Y and YZ Aurigæ, WY Tauri and RX Andromedæ. Williams recognized the inherent risk of missing variables that had periods on the order of one day. He chose to deal with that by making multiple exposures during one evening. It was this technique that allowed him to discover several eclipsing variables of very short period. Williams was the first English amateur to apply photography in this manner, although variable stars had been discovered on photographic plates at the Harvard College Observatory for several years. [15]

Analysis and Theoretical Interpretation of Variable Star Observations

The final area of amateur contributions to variable star astronomy to be considered in this paper is the collection, "Analysis and Theoretical Interpretation of Variable Star Observations". In this area the identifiable contributions are more sparse than the previous two areas. Generally speaking, amateur astronomers come to the field less prepared academically for this work. Nonetheless, there are a few important contributors to discuss, some of which are listed in Table 3 (see end of text).

M. A. Blagg

Mary Adela Blagg (1858-1944) was attracted to variable star astronomy after making very noteworthy contributions to Selenography. At the conclusion of her lunar work, H. H. Turner induced her to take up the reduction and analysis of the variable star observations of Joseph Baxendell. The results were published in a series of ten papers in the Monthly Notices of the Royal Astronomical Society. When this work was completed Ms. Blagg undertook studies of the observational data available for a number of other stars including β Lyræ, RT Cygni, V Cassiopeiæ and U Persei. In each case she reduced the data to define new elements for the star, and harmonically analyzed the resulting light curves. Her work displays "skill and sound judgement,.... originality and courage". [16]

F. de Roy

One of the leading Belgian variable star astronomers was Félix Eugène Marie de Roy (1883-1942). A journalist by profession, de Roy was very involved in the organization and popularization of astronomy in Belgium as well as with the observation of meteors. He was already well known to variable star astronomers by 1921 when AAVSO Recorder Leon Campbell proposed that he be elected to membership in the AAVSO. The application form for this election notes that de Roy was, at that time, President of the Societe d' Astronomie Belgium. The following year, de Roy was elected Director of the BAA-VSS replacing C. L. Brock. In 1928 the AAVSO Council elected de Roy an Honorary Member, recognizing his invaluable services to variable star astronomy as Director of the BAA-VSS.

Analysis of observations captured de Roy's imagination, and he devoted substantial effort to this activity. For example, in 1932 de Roy proposed, after exhaustive analysis, that RX Andromedæ and Z Camelopardalis should not be classified with U Geminorum and SS Cygni stars but should be considered for a separate classification of their own. The University of Utrecht conferred the degree Doctor Honoris Causa upon de Roy in 1936, recognizing his exemplary services to astronomy in both variable stars and in meteors. [17]

E. E. Markwick

Much of the early progress of the BAA-VSS is traceable to the profound influence of one individual, Ernest Elliott Markwick (1853-1925). The first Director of the VSS, John Ellard Gore, had been diligent in publishing the variable star observations of individual members. But Gore never devoted effort to the consolidation of observations, analysis of the consolidated results or organizing and expanding the efforts of individual observers. It was in these areas that Markwick made his greatest contributions, and in fact set the pattern of activity followed by the BAA-VSS since then.

Colonel Markwick had a distinguished military career. He was commissioned in the Army Ordnance Department and saw service in a number of active conflicts. His sense of organization and discipline from these experiences were, no doubt, an important influence in his contributions to astronomy. In 1893 the Army assigned Markwick to a post in Gibraltar. With his new exposure to the southern skies, Markwick undertook a survey to discover variable stars. He was successful in his endeavor, and is credited with the discovery of T Centauri and RY Sagittarii. Discovery of either star was very difficult using purely visual techniques.

However, it is Markwick's contributions in reorganizing the efforts of the BAA-VSS and initiating analysis of the section's observational results to which we now wish to call attention. Markwick returned to England from Gibraltar in 1898. Within a year he was asked to take over the VSS Director's post from Gore. He immediately restructured the activities of the VSS by 1) narrowing the work list to 46 stars which all VSS observers were encouraged to observe, 2) adopting uniform magnitudes for comparison stars, and 3) adopting a uniform method of observing and making estimates of the variable star's brightness and a standard form for reporting results.

The changes Markwick initiated went well beyond these procedural aspects. He also established a much needed communication network for the VSS members with frequent letters, circulars, reports and memoirs. Markwick also undertook the compositing of the results from various observers and preparation of a consolidated analysis and report for each star on a regular basis. Finally, Markwick wrote an extended series titled "Variable Stars and How to Observe Them" which appeared in the English Mechanic. The members of the VSS responded with new enthusiasm and commitment to variable star astronomy. It is no small tribute to Markwick's foresight that many of the changes he introduced during his ten years as the VSS Director have remained in place over the intervening years. [18]

A. W. Roberts

Alexander William Roberts (1858-1938) was born and educated in Scotland. In 1883, he migrated to South Africa to devote a lifetime to the training of native teachers at Lovedale in the Cape Colony. Roberts earned a reputation as both an outstanding teacher and as a humanitarian friend of the natives.

Roberts had developed an early interest in astronomy in Scotland. By 1888 his interest had shifted specifically to variable stars, and in 1891 he began a systematic survey of the sky. Using only small binoculars and an old 1-inch theodolite, Roberts discovered no less than twenty new variable stars in the next three years. Four of these discoveries were Algol type eclipsing binary stars.

Roberts' remarkable early achievements with very limited equipment drew substantial attention to his efforts. In 1899, David Gill, Director of the Cape Observatory, arranged for the acquisition of a specially designed 2-inch prismatic photometer. This instrument allowed Roberts to achieve very high levels of precision in magnitude estimates. The resulting light curves proved susceptible to elegant analysis which would not have been possible with more scatter in the data.

In the best traditions of science Roberts sought to find some meaning in the observations he was making. With his high precision light curves Roberts was able to extend his work into theoretical understanding of the stars he observed. To cite two examples:

1. Roberts was the first astronomer to calculate the upper limits of densities of individual stars in close binary systems and show that they were extremely tenuous objects.

2. Roberts was the first to give observational proof that stars in close binaries are distorted by tidal effects and to demonstrate that the resulting shapes are prolate spheroids in agreement with the then recent theories of Darwin and Poincaré.

Roberts' work on close binaries is widely recognized and noted in the literature on these systems. In large measure, this recognition came about because Roberts dedicated the time and energy to interpret his own observations from a theoretical viewpoint and then prepare the results of his work for publication. For it is only through the publication of 70 papers in various journals that Roberts was able to make known to the scientific world the results of his work on the veld of South Africa. With these substantial contributions--observational technique and results, discovery and theoretical analysis-- Alexander William Roberts stands out as one of the truly great amateur contributors to variable star astronomy.[19]

Summary - Amateur Contributions to Variable Star Astronomy

The intent of this paper is to call attention to a few of the many noteworthy variable star astronomers from around the world. From the individual cases described here it is apparent that there are many different circumstances under which amateurs make substantial contributions to variable star astronomy. Over the past 100 years amateur astronomers have provided a sizeable portion of the raw material for each new phase of variable star astronomy, whether it be in the discovery of new variable stars or in the development of long term observational records. Amateurs even make contributions to the analysis and theoretical understanding of the data they create. Although that contribution has been limited in the past, with the advent of personal computers, and with access to central data bases provided by telephone modems, there is a great opportunity for analysis and theoretical contribution by amateurs in the future.

Clearly, as in the past, and likely far into the future, the opportunity is there for amateur astronomers with limited resources to make a greater contribution to astronomy in the study of variable stars than in any other astronomical area.

Discussion

Halbach: Let's find a new name for us "observing amateur astronomers", contrasted from "armchair" astronomers.

Williams: My own belief is that most such individuals are not amateur astronomers in the first place. I choose to coin a new name for them - *Recreational Sky Observers*, and reserve the name *Amateur Astronomer* for those who practice astronomy as a science.

End Notes

Space limitations have made it necessary to limit references to only one for each of the astronomers discussed in this paper.

1. Rosebrugh, David W., "Thirty-Fifth Annual Meeting of the A.A.V.S.O. Harvard Observatory, Cambridge, Massachusetts, October 11 and 12, 1946", Variable Comments, IV, 15 (1947?), 76.
2. Williams, T. R., "Criteria for Identifying an Astronomer as an Amateur", in Dunlop, S. and M. Gerbaldi (Eds), Stargazers, The Contributions of Amateurs to Astronomy, Proceedings of Colloquium 98 of the IAU, Berlin Heidelberg: Springer-Verlag, 1988.
3. Williams, Thomas R., "Three AAVSO Leaders: de Kock, Fernald and Peltier", JAAVSO, 15, 2 (1986), 135.
4. Lindley, W. M., "Obituary Notices - Charles Frederick Butterworth", MNRAS, 107, 1 (1947), 40.
5. Hoffleit, Dorrit, "News Notes", Sky and Telescope, 2 (24), 14; and personal communication from Marie Peddle to Thomas R. Williams dated November 3, 1986.
6. Forbes, A.F.I. "Obituary - Hendon Egerton Houghton, M.B.E., F.R.A.S.", MNASSA, 1, 1 (1948) and personal communication from Marie Peddle to Thomas R. Williams dated November 3, 1986.
7. Chandra, Radha Govinda, Application forms for membership in AAVSO dated 29 January 1920 and 29 September 1920.
8. Favero, Giancarlo and Sandro Baroni, "Giovanni Battista Lacchini: An Amateur Astronomer from Italy", in Stargazers, The Contributions of Amateurs to Astronomy, Proceedings of Colloquium 98 of the IAU, June 20-24, 1987, S. Dunlop and M. Gerbaldi (Eds.), Berlin: Springer-Verlag, 1988.
9. Hurless, Carolyn, "We Remember Curtis...", JAAVSO, 6, 1 (1977), 9.
10. Mattei, Janet A., "Carolyn Hurless Obituary", manuscript included in personal communication to Tom Williams, letter dated 4/3/1987.

11. Vargha, Mrs. M., "Hungarian Astronomy of the Era", in A Kalocsai Haynald Obszervatórium Története, Dr. Mojzes Imre Ed., Budapest (1986).
12. MacPherson Jr., Hector, "Two Scottish Astronomers of Today", Popular Astronomy, XVI, 7 (1908 August-September), 397.
13. Maunder, E. E., "Obituaries - The late T. H. Astbury. An Appreciation", JBAA, 34, 8 (1924 June), 327.
14. Minois, Joel, "Antoine Brun: Founder of the AFOEV", JAAVSO, 16, 2 (1987), 143.
15. Phillips, T. E. R., "Arthur Stanley Williams", MNRAS, 99, 4 (1939 February), 313.
16. Ryves, P. M., "Obituary Notices - Mary Adela Blagg", MNRAS, 105, 2 (1945 April), 65.
17. Obituary Note "M. Felix De Roy", Gazette Astronomique, 28, 1, 2, 3 (1946 January-March), 3-4.
18. Anonymous Obituary, "Colonel Ernest Elliott Markwick, C.B., C.B.E., late Army Ordnance Department", JBAA, 35, 9 (1924-25), 311.
19. McIntyre, Donald G., Obituary. Alexander William Roberts (1857-1938), JASSA, IV, 3 (1938 April), 117.

Table 1

VARIABLE STAR OBSERVERS

Name	Dates	Country
Ahnert, Paul	1897-1989	Germany
Anderson, Curtis E.	1927-1977	USA
Anderson, Thomas David	1853-1932	England
Aoki, Masahiro	1920-1984	Japan
Armfield, LaVerne E.	1906-	USA
Astbury, Thomas Hinsley	1858-1922	England
Bancroft, H. C. Jr.	1884-	USA
Baxendell Sr., Joseph	1815-1887	England
Beyer, Max	1894-1982	Germany
Bicknell, R. H.	-	Rhodesia
Birmingham, John	1816-1884	Ireland
Birt, William Radcliff	1804-1881	England
Boss, Lewis Judson	1898-1982	USA
Bouton, Rev. Tilton Clark Hall	1856-1948	USA
Brocchi, Dalmero F.	1871-1955	USA
Brun, Antoine	1881-1978	France
Buckstaff, Ralph N.	1887-1980	USA
Butterworth, Charles Frederick	1870-1946	England
Chambers, George Frederick	1840-1915	England
Chandra, Rhada Govinda	1881-	India
Chassapis, Constantin	-	Greece
Cilley, Morgan	1878-1955	USA
Crust, Alec G.C.	-1945	New Zealand
De Roy, Felix	1883-1942	Belgium
DeKock, Reginald Purdon	1902-1980	South Africa
Duruy, Maurice Victor	1894-1984	France
Elias, Demetrius P.	1926-	Greece
Engelkemeir, Donald W.	1920-1969	USA
Ensor, George Edmund	1873-1943	South Africa
Evershed, Mary Acworth Orr	1867-1949	England
Fernald, Cyrus F.	1901-1979	USA
Geddes, Murry	1909-1944	New Zealand
Goldschmidt, Hermann	1802-1866	France
Gore, John Ellard	1845-1910	Ireland
Gow, John Graham	1884-1960	New Zealand
Hartmann, Ferdinand	1899-1965	USA
Hoffmeister, Cuno	1892-1968	Germany
Holborn, Frank Maurice	1884-1962	England
Holt, William L.	1879-1946	USA
Honda, Hideo	1920-1983	Japan
Houghton, Hendon Egerton	1892-1947	South Africa

Noteworthy Variable Star Astronomers - continued

Hurless, Carolyn J.	1934-1987	USA
Jones, Eugene H.	1864-	USA
Kanda, Shigeru	1894-1974	Japan
Kasai, Yoshihiko	1903-1961	Japan
Kearons, Winifred Crosland	-1958	USA
Kirchhoff, Peter	1893-1976	South Africa
Kozawa, Kiichi	1918-1949	Japan
Kruytbosch, W. E.	-1934	Netherlands
Köhl, Torvald	1852-1931	Denmark
Lacchini, Giovanni Battista	1884-1967	Italy
Laskowski, Zygmunt	1841-1928	Poland
Loreta, Guiseppe	1909-1945	Italy
Markwick, Ernest Elliot	1853-1925	England
Masterman, Stillman	1831-1863	USA
McAteer, Charles Y.	1865-1924	USA
Meek, Joseph W.	1912-	USA
Meesters, P. G.	1887-1964	Netherlands
Miyajima, Zen-ichiroh	1887-1946	Japan
Morgan, Francis P.	1906-1976	Canada
Morley, Leonard George Edward	1894-1973	New Zealand
Morrisby, Arthur Gustave Fairfax	1926-1988	Zimbabwe
Naef, Robert A.	1907-1975	Switzerland
Okabayashi, Shigeki	1913-1944	Japan
Olcott, William Tyler	1873-1936	USA
Parkhurst, Henry M.	1825-1908	USA
Parkhurst, John Adelbert	1861-1925	USA
Peltier, Leslie C.	1900-1980	USA
Penrose, Francis C.	1818-1903	England
Phillips, Theodore Evelyn Reece	1868-1942	England
Pickering, David Bedell	1873-1946	USA
Pigott, Edward	1753-1825	England
Podmaniczky, Geza	1839-1923	Hungary
Pogson, Norman Robert	1829-1891	England
Posztoczky, Károly	1882-1963	Hungary
Prentice, Edward	1864-1924	England
Roberts, Alexander William	1857-1938	South Africa
Rosebrugh, David W.	-1988	USA
Safarik, Adalbert	1826-1902	Czechoslovakia
Saw, Douglas R. B.	-1990	England
Sawyer, Edwin F.	1849-	USA
Shinkfield, Reginald Claxton	1901-1984	Australia
Skjellerup, John Francis	1875-1952	South Africa
Steavenson, William Herbert	1894-1975	England

Noteworthy Variable Star Astronomers - continued

Szeligiewicz, Edward	1924-1959	Poland
Taboada, Domingo	1893-	Mexico
Topham, Bertram John	1894-	Canada
Venter, Stephanus Christiaan	1907-1968	South Africa
Webb, Harold B.	1890-	USA
Williams, Arthur Stanley	1861-1938	England
Yendell, Paul S.	1845-1918	USA

Table 2

VARIABLE STAR DISCOVERERS

Abbott, Francis	1799-1883	Australia
Anderson, Thomas David	1853-1932	England
Astbury, Thomas Hinsley	1858-1922	England
Baxendell Sr., Joseph	1815-1887	England
Birmingham, John	1816-1884	Ireland
Brun, Antoine	1881-1978	France
Chilton, Kenneth Edward	1939-1976	Canada
Crust, Alec G.C.	-1945	New Zealand
DeKock, Reginald Purdon	1902-1980	South Africa
Degenfeld-Schomberg, Berta	1884-1928	Hungary
Espin, Thomas Henry Espinall Compton	1858-1934	England
Fernald, Cyrus F.	1901-1979	USA
Gore, John Ellard	1845-1910	Ireland
Hind, John Russell	1823-1895	Enqland
Hoffmeister, Cuno	1892-1968	Germany
Honda, Hideo	1920-1983	Japan
Hudson, George Vernon	1867-1946	New Zealand
Jonckheere, Robert	1888-1974	Belgium, France
Kirchhoff, Peter	1893-1976	South Africa
Komáromi-Kacz, Endre	1880-1968	Hungary
Lacchini, Giovanni Battista	1884-1967	Italy
Laskowski, Zygmunt	1841-1928	Poland
Markwick, Ernest Elliot	1853-1925	England
Masterman, Stillman	1831-1863	USA
Metcalf, Joel Hastings	1866-1925	USA
Okabayashi, Shigeki	1913-1944	Japan
Peltier, Leslie C.	1900-1980	USA
Pigott, Edward	1753-1825	England
Pogson, Norman Robert	1829-1891	England
Prentice, John Phillip Manning	1903-1981	England

Noteworth Variable Star Astronomers - continued

Rhorer, S. Lynn	1870-1929	USA
Roberts, Alexander William	1857-1938	South Africa
Saw, Douglas R. B.	-1990	England
Steavenson, William Herbert	1894-1975	England
Williams, Arthur Stanley	1861-1938	England
Yendell, Paul S.	1845-1918	USA

Table 3

ANALYSIS AND THEORY OF VARIABLE STARS

Barr, Joseph Miller	1857-1911	Canada
Birt, William Radcliff	1804-1881	England
Blagg, Mary Adela	1858-1944	England
De Roy, Felix	1883-1942	Belgium
Gore, John Ellard	1845-1910	Ireland
Holborn, Frank Maurice	1884-1962	England
Markwick, Ernest Elliot	1853-1925	England
Monck, William Henry Stanley	1839-1915	Ireland
Parkhurst, John Adelbert	1861-1925	USA
Phillips, Theodore Evelyn Reece	1868-1942	England
Pogson, Norman Robert	1829-1891	England
Prentice, Edward	1864-1924	England
Roberts, Alexander William	1857-1938	South Africa
Saw, Douglas R. B.	-1990	England
Williams, Arthur Stanley	1861-1938	England
Yendell, Paul S.	1845-1918	USA

AAVSO AND VARIABLE STAR OBSERVING

Janet Akyuz Mattei
AAVSO
25 Birch Street
Cambridge, Massachusetts 02138
USA

1. AAVSO

The American Association of Variable Star Observers (AAVSO) was founded in 1911 at Harvard College Observatory to coordinate variable star observations made largely by amateur astronomers and to make them available to professional astronomers. In 1954, the AAVSO became an independent, private, nonprofit research organization. Today with members in 43 countries, and headquarters in Cambridge, Massachusetts, it is the world's largest association of variable star observers.

Membership in the AAVSO is open to anyone interested in variable stars and in contributing to the support of valuable research. Research on variable stars is important because they can provide much information about stellar properties and evolution, which can then be extrapolated to other types of stars. Data about mass, radius, internal and external structure, composition, temperature, luminosity, and distance can be gained by the study of variable stars. Since professional astronomers have neither the time nor the telescopes needed to gather data on the brightness changes of thousands of variables, amateurs can make valuable contributions to this field of science by observing these stars.

The AAVSO coordinates, evaluates, compiles, processes, publishes, and disseminates observations to the astronomical community throughout the world. The scientific activities of the AAVSO are coordinated by the Director of the Association, who is a professional astronomer.

The archives of the AAVSO currently contain over 6.5 million visual observations and about 5,000 photoelectric ones. Approximately 550 observers from around the world send in between 240,000 and 265,000 observations yearly. A person does not need to be an AAVSO member in order to submit observations to the AAVSO. At the end of each month, incoming observations are sorted by observer and checked for errors. They are then converted into computer-readable form and processed using computer systems at AAVSO Headquarters. These observations are then added to the data files for each star, and the corresponding computer-generated light curves are brought up to date.

2. AAVSO Observing Programs

The AAVSO Visual Observing Program contains 3300 variable stars. Stars best suited for visual observing have amplitudes of variations more than one magnitude. Therefore pulsating variables (long period, semiregular, irregular, R Coronae Borealis, RV Tauri, Cepheid, RR Lyrae types), eruptive variables (nova, dwarf nova, recurrent nova, nova-like, symbiotic types) nebular and flare stars, quasars with optical variability, and eclipsing binaries make up the visual observing program. The accuracy of the visual observations is between ± 0.2 and ± 0.4 magnitude and the limiting magnitude is about 16.5.

The AAVSO Photoelectric Observing Program contains 50 bright, red variables with a small amplitude of variation, such as semiregular, symbiotic, RV Tauri, R Coronae Borealis, and irregular types of variables. Most of these stars are also in the Visual Observing Program, as they also show long-term, large-amplitude variations. Thus, the photoelectric and the visual data on these stars complement each other. The accuracy of photoelectric observations is ± 0.008 magnitude.

AAVSO finder charts identifying the variables and comparison stars of known magnitudes are available for most stars in its programs. The magnitudes of the comparison stars have been determined photoelectrically, photovisually using an iris photometer, or, in some cases, visually. All observers are asked to use the same charts when making observations they plan to submit to the AAVSO. Their following this procedure is essential for standardization of reported magnitude estimates.

3. Naming of Variable Stars

Each variable is identified by both a name and numerical designation. Names are assigned in the order in which the variable stars are discovered in a constellation. The first variable is given the letter R, the next S, and so on to Z. The next star is named RR, then RS, etc. to RZ; SS to SZ, and so on to ZZ. Then naming starts over at the beginning of the alphabet: AA, AB and so on to QZ. This system (omitting the letter J) can accommodate 334 names. There are so many variables in some constellations in the Milky Way, however, that an additional nomenclature is necessary. After QZ, variables are named V335, V336, etc. These names are then combined with the genitive form of the constellation name, as in SS Cygni or V378 Scorpii.

The numeral designation (called the Harvard designation because of its origin) is a group of six numbers that give the variable's approximate coordinates for the year 1900.0. The first four digits give the hour and minutes of right ascension, the last two the degrees of declination. (Traditionally, if the variable is in the southern hemisphere, the last two digits are underlined or printed in either italic or boldface type.) Thus the designation 094211 for R Leonis denotes an approximate 1900 right ascension of 09h 42m and a declination of $+11°$. The designation 1702<u>15</u> for R Ophiuchi indicates 17h 02m, -15°. With the advent of computers, it has become convenient to put + and - signs in front of the last two digits; thus, the designations for R Leonis and R Ophiuchi become 0942+11 and 1702-15, respectively.

4. Types of Variable Stars

Variable stars are classified as either intrinsic, wherein variability is caused by physical changes such as pulsation or eruption in the star or stellar system, or extrinsic, wherein variability is caused by the eclipse of one star by another or by the effects of stellar rotation.

4.1. Pulsating Variables

Cepheids: These variables pulsate with periods from 1 to 70 days, with an amplitude of light variation from 0.1 to 2 magnitudes. They have high luminosity and are of F spectral class at maximum and G to K at minimum. The later the spectral class of a Cepheid, the longer is its period. Cepheids obey the well-known period-luminosity relation. Example: delta Cephei.

RR Lyrae stars: These short-period, pulsating variables have periods ranging from 0.05 to 1.2 days and light amplitudes between 0.3 and 2 magnitudes. They are white giant stars, usually of spectral class A. Example: RR Lyrae.

RV Tauri stars: These yellow supergiants have a characteristic light variation with alternating deep and shallow minima. Their periods, defined as the interval between two deep minima, range from 30 to 150 days. The light amplitude may be as much as three magnitudes. Some show long-term cyclic variations ranging from hundreds to thousands of days. Generally, the spectral class ranges from G to K. Example: R Scuti (Figure 1).

Long period or Mira variables: These giant red variables vary with periods ranging from 80 to 1000 days and visual light amplitudes from 2.5 to 11 magnitudes or more. They have characteristic late type emission spectra - Me, Ce, Se. Example: chi Cygni (Figure 2).

Semiregular variables: These giants and supergiants of intermediate and late type spectral types show appreciable periodicity accompanied by intervals of irregular light variation. Their periods range from 30 to 1000 days, with light amplitudes of not more than one to two magnitudes. Example: V Bootis (Figure 3).

4.2. Eruptive Variables

Supernovae: These show brightness increases of 20 or more magnitudes as a result of a catastrophic stellar explosion. Example: CM Tauri (Supernova of A.D. 1054 and the central star of the Crab Nebula).

Novae: These close binary systems with periods from 0.05 to 230 days, consist of a giant, subgiant, or a dwarf star of K to M spectral class as a cool component and a dwarf star as a primary, hot component which increases in brightness by 7 to 19 magnitudes (V) in a matter of one to several hundred days. After the outburst, the star slowly fades to its initial brightness over several years or decades. Near maximum brightness, the spectra are generally similar to absorption spectra A or F giant stars. Example: V1668 Cygni (Figure 4).

Recurrent novae: These objects are similar to novae, but they have had two or more outbursts during their recorded history. The light curves for each of the eruptions are very similar. Example: RS Ophiuchi (Figure 5).

U Geminorum stars: These are close binary systems consisting of a dwarf or subgiant star of K to M spectra and a hot white dwarf surrounded by an accretion disk and with orbital periods from 0.05 to 0.5 days. After intervals of quiescence at minimum light, they brighten suddenly. Depending on the star, the eruptions occur at intervals of 10 to thousands of days. The light amplitude of the outbursts ranges from 2 to 6 magnitudes, the duration from 5 to 20 days. Example: SS Cygni (Figure 6).

Z Camelopardalis stars: These stars are physically and spectroscopically similar to U Geminorum stars. They show cyclic variations interrupted by intervals of constant brightness (stillstands), approximately a third of the way from maximum to minimum and lasting from tens to hundreds of days. Example: Z Camelopardalis (Figure 7).

SU Ursae Majoris stars: These stars are also physically and spectroscopically similar to U Geminorum stars. They have two distinct kinds of outbursts: one is faint, frequent, and short, with a duration of one to two days; the other (supermaximum) is bright, less frequent, and long, with a duration of 10 to 20 days. During supermaxima, small-amplitude, periodic modulations (superhumps) appear with periods two to three percent longer than the orbital period of the system. Example: AY Lyrae (Figure 8).

Symbiotic stars: These close binary systems have as one component a red giant, the other a hot blue star, and both embedded in nebulosity. They show semi-periodic, nova-like outbursts up to three magnitudes in amplitude. Example: Z Andromedae (Figure 9).

R Coronae Borealis stars: These luminous, hydrogen-poor, carbon- and helium-rich supergiants, spend most of their time at maximum light and at irregular intervals fade by 1 to 9 magnitudes, then slowly recover to their normal brightness after a few months to a year. Superimposed on these variations are periodic pulsations of several tenths of a magnitude and periods ranging from 30 to 100 days. Members of this group are of F to K and R spectral types. Example: R Coronae Borealis (Figure 10).

4.3. Eclipsing Binaries

These are binary systems of stars with an orbital plane lying near the line of sight of the observer. The components periodically eclipse each other, causing a decrease in the apparent brightness of the system as seen by the observer. The period of the eclipse, which coincides with the orbital period of the system, can range from minutes to years. Example: beta Persei (Algol).

4.4. Rotating Variables

These are rotating stars, often binary systems, which undergo small-amplitude changes in light that may be due to dark or bright spots or patches on their stellar surface (starspots). Example: RS Canum Venaticorum.

5. Establishing an Observing Program

To obtain the maximum benefits from variable star observing, such as scientifically useful data and personal satisfaction, observers should follow an observing program that is suited to their interests, experience, equipment, and observing site and conditions.

Geographic location: The scale of the observing program will be influenced by the location and the terrain of the observer's site.

Sky conditions: The annual percentage of clear nights influences the selection of types of variables. The larger the percentage of clear nights, the more advisable it is to go after stars that require nightly observations, such as the cataclysmic variables and R Coronae Borealis stars. If an observer has only 20 percent or fewer clear nights per year, however, it is recommended that slowly-varying long period variables be selected, since for these stars even one observation a month is meaningful.

Light pollution: Light pollution affects the limiting magnitude of observations. Thus, an observer living in a city is advised to concentrate on observing bright stars well above the limiting magnitude of the instrument, while observers with dark skies should be challenged to go after stars as faint as their instruments will allow.

Optical instrument: Successful variable star observing requires interest, perseverance, and the proper optical instrument. A good pair of binoculars is sufficient for bright stars, while for faint stars one needs either a small or large reflector or refractor telescope, either portable or permanently mounted. There is no "ideal" telescope for variable star observing; each has its own special advantage. More important is familiarity with the instrument at hand and a determination to accomplish as much as possible with it.

The most popular type of telescope among variable star observers is a short-focus (f/5 - f/8) Newtonian reflector, with aperture six inches or more, equipped with a good finder. Such an instrument is easy to build or inexpensive to buy, and more important, it is convenient to use and easy to maneuver.

A low power wide-field eyepiece is an important observing aid in locating variable stars, and it allows the observer to include as many of the comparison stars in the field as possible. High magnification is not necessary unless one is observing faint stars or congested fields. It is important to use the best quality eyepieces that one can afford. Wide-field and Plossl types are among the best. A good quality, achromatic, two-power Barlow lens is also a valuable aid.

Observing program: Beginning visual observers usually start with long period and semiregular variables. Their large amplitudes and slow rates of variation make their observation an ideal means of observers' becoming experienced at making brightness estimates.

As observers gain experience, they often add to their programs other types of variable stars - cataclysmic variables like SS Cygni, U Geminorum, and Z Camelopardalis; symbiotic stars like Z Andromedae; and R Coronae Borealis stars. These stars vary unpredictably, and depending on their state of behavior, may require very frequent observations.

Some observers include or focus on eclipsing binaries or RR Lyrae stars. Observing these stars requires additional preplanning and special observing techniques.

Observers may also include in their programs systematic, regular searches for novae or supernovae.

Circumpolar variables, which can be observed all year long, are very worthwhile observing projects for observers appropriately located. Also, experienced observers may wish to make observations in evening or morning twilight, as observations made at these times are particularly valuable.

6. Services to the Observer

The AAVSO enables variable star observers to contribute vitally to variable star astronomy by accepting their observations, incorporating them into the AAVSO data files, publishing them, and making them available to the professional astronomer. Incorporating an observer's observations into the AAVSO archives means that future researchers will have access to those observations, so the observer is contributing to the science of the future as well as the present. Also, the AAVSO coordinates observing runs between professional and amateur astronomers, in which observations from amateur astronomers play an important role in correlating observations obtained with special instruments at earth-based observatories or aboard satellites.

On request, the AAVSO will help set up an appropriate observing program for an individual, an astronomy club, an elementary school, high school, or college, etc. In this way observers, students, and faculty are able to make the best use of their resources and to do valuable science. The AAVSO can also assist in teaching observing techniques and in suggesting stars to be included in a program.

7. Services to the Astronomical Community

AAVSO data, both published and not-yet-published, are disseminated extensively to astronomers around the world, upon request. AAVSO services are sought by astronomers for the following purposes:

a) Real-time, up-to-date information on unusual stellar activity;

b) Assistance in scheduling and executing of variable star observing programs using earth-based large telescopes and instruments aboard satellites;

c) Assistance in simultaneous optical observations of program stars and immediate notification of their activity during earth-based or satellite observing programs;

d) Correlation of AAVSO optical data with spectroscopic, photometric, and polarimetric multi-wavelength data;

e) Collaborative statistical analysis of stellar behavior using long-term AAVSO data.

Collaboration between the AAVSO and professional astronomers for real-time information or simultaneous optical observations has enabled the successful execution of many observing programs, particularly those using satellites such as Apollo-Soyuz, High Energy Astronomical Observatories 1 and 2 (HEAO 1 and 2), International Ultraviolet Explorer (IUE), the European X-Ray Satellite (EXOSAT), and High Precision Parallax Collecting Satellite (HIPPARCOS). A significant number of rare events have been observed in multi-wavelength with these satellites as a result of timely notification by the AAVSO.

AAVSO visual and photoelectric data and services may be obtained by writing to the Director and stating the purpose of the request and the type of AAVSO data or services needed.

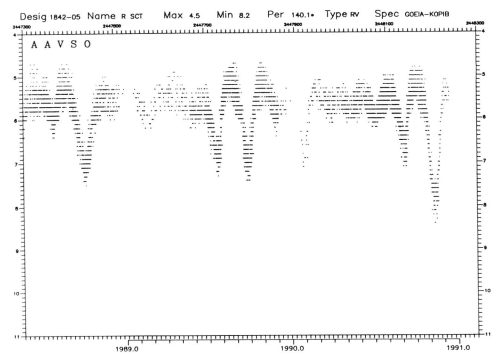

Figure 1. AAVSO visual light curve of R Scuti.

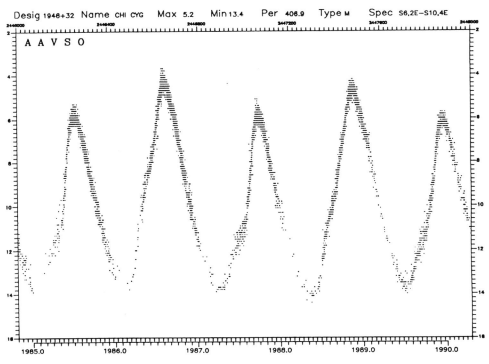

Figure 2. AAVSO visual light curve of chi Cygni.

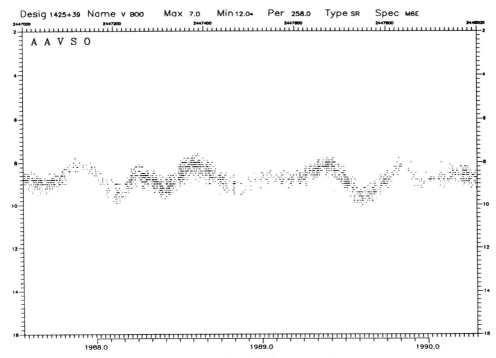

Figure 3. AAVSO visual light curve of V Bootis.

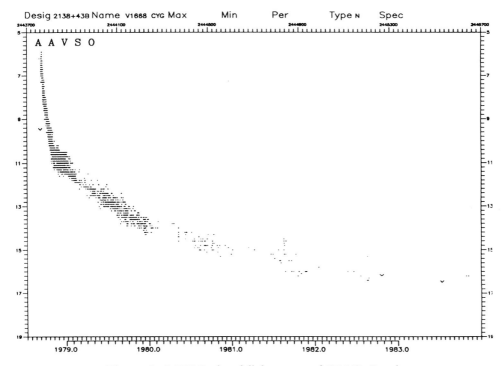

Figure 4. AAVSO visual light curve of V1668 Cygni.

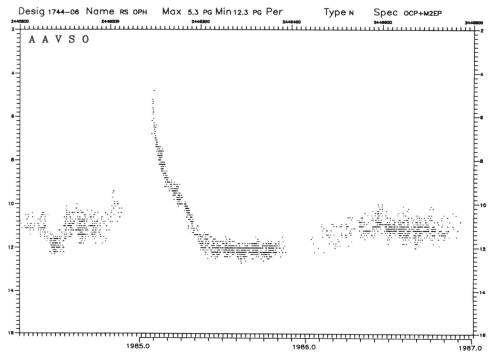

Figure 5. AAVSO visual light curve of RS Ophiuchi.

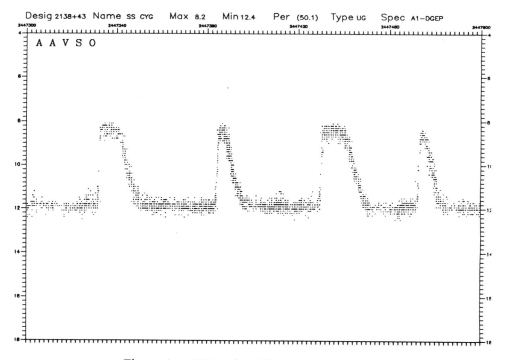

Figure 6. AAVSO visual light curve of SS Cygni.

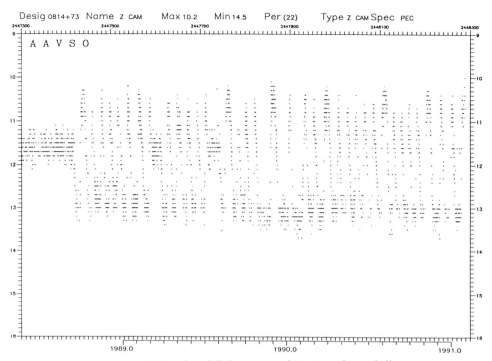

Figure 7. AAVSO visual light curve of Z Camelopardalis.

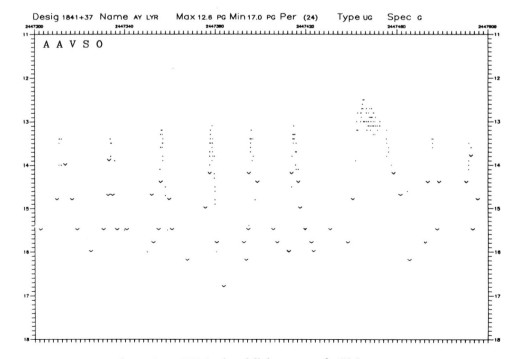

Figure 8. AAVSO visual light curve of AY Lyrae.

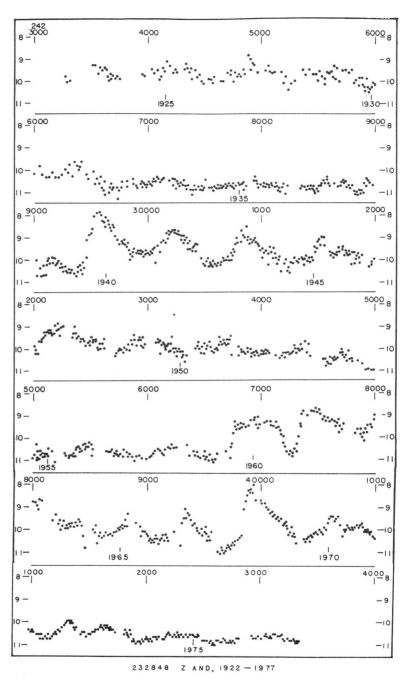

Figure 9. *AAVSO long-term visual light curve of Z Andromedae.*

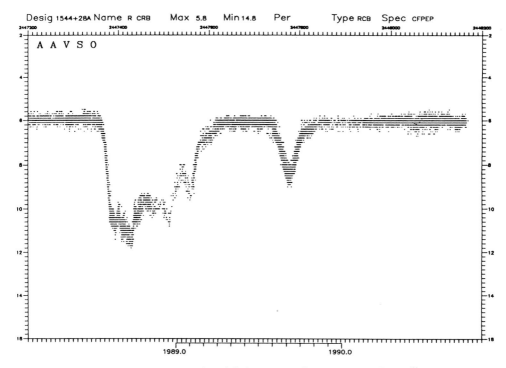

Figure 10. *AAVSO visual light curve of R Coronae Borealis.*

INTERNATIONAL ASTRONOMICAL UNION
COMMISSION 27: VARIABLE STARS

Michel Breger
Institute of Astronomy/University of Vienna
Türkenschanzstrasse 17/ A-1180 Wien, Austria

The International Astronomical Union (IAU) was founded in 1919 to provide a forum where astronomers from all over the world can develop astronomy in all its aspects through international cooperation. The Union has Adhering Member Organizations in 56 countries and about 6700 individual members. The IAU, among other things, is the sole authority for assigning designations and names to celestial bodies and the surface features on them. Over 330 major scientific meetings have been sponsored by the IAU. The scientific activity is also reflected in the work of its 40 commissions.

The members of Commission 27 (Variable Stars) are mostly professional astronomers active in the field with several publications in refereed journals. These formal requirements for IAU membership are set by the national committees. In addition, the commissions may propose new members who have made exceptional contributions to astronomy. Naturally, the distinction between professional and amateur astronomers is not absolute.

Commission 27 provides several services to the astronomical community:

(i) The commission sponsors the publication of the *Information Bulletin on Variable Stars*. An enormous amount of work has been put in by the editors Bela Szeidl and Laszlo Szabodos at Konkoly Observatory, Hungary. Over 3500 issues describing new discoveries have been published so far. A limited number of copies is printed and sent to libraries and many commission members.

(ii) The commission maintains the *Archives of Unpublished Photoelectric Observations of Variable Stars* (coordinator Michel Breger) with branches in Strasbourg (France), London (England), and Odessa (USSR). Data are stored in paper and electronic form. So far over 220 files containing between 1 and 71 stars each have been assigned. The new files are announced every few years in the *Publications of the Astronomical Society of the Pacific* as well as the *Bulletin Centre de donnees stellaires* in Strasbourg, which also issues cross-references of the star names in the Archives.

(iii) The commission sponsors the important *General Catalog of Variable Stars* (Editor-in-chief N. N. Samus), which is published by the Astronomical Council of the USSR Academy of Sciences and updated periodically. This catalog is essential for all workers in the field of variable stars.

(iv) Every three years the commission issues a triannual report on the new developments in the field of variable stars. Over a dozen selected experts summarize the latest work providing probably the best astronomical reviews available anywhere. These 36 pages filled with news and references are published in the *Transactions of the IAU*. We will complete a new report by September, 1990 (to be published a year later). Preprints can be sent to interested individuals.

The IAU holds a General Assembly every three years with the next meeting scheduled for July, 1991 in Buenos Aires, Argentina. Commission 27 will organize several scientific sessions at this General Assembly, including a session on the *Role of Rotation on Stellar Variability*.

The AAVSO is well represented in Commission 27. My duties as president of the commission will be completed during 1991 and the new president will probably be John Percy, the present vice-president of the commission and current president of the AAVSO.

THE PRESENT STATE OF THE COMPILATION OF THE GENERAL CATALOGUE OF VARIABLE STARS (GCVS)

N. Samus
The Astronomical Council of the USSR Academy of Science
48, Pyatnitskaya Str.
Moscow 109017, USSR

The history of variable star catalogues starts as early as in the 18th century. In our century, a catalogue of variable stars (with ephemerides) was published every year from 1926 till 1935 by R. Prager and from 1936 till 1943 by H Schneller on behalf of the German *Astronomische Gesellschaft*. In 1946, the Executive Committee of the IAU decided to continue publishing variable star catalogues in the Soviet Union. This decision was based on the existence of a good variable star card catalogue in Moscow.

The Soviet period of the catalogue is connected with the activity of such outstanding astronomers as P. P. Parenago (1906-1960), B. V. Kukarkin (1909-1977), P.N. Kholopov (1922-1988). The first three editions of the GCVS are characterized by the following information.

GCVS I	Single volume	10930 stars	1947
GCVS II	Two volumes	14708 stars	1958
GCVS III	Three volumes	20437 stars	1969-1971

Before we started our work on the 4th GCVS edition, we prepared the *New Catalogue of Suspected Variable Stars* (NSV, 14810 stars, 1982). By the way, we insist on using the abbreviation "NSV" for the stars of this catalogue; those using the "SVS" (Suspected Variable Star) abbreviation should remember that there already exist preliminary designations of Soviet-discovered variable stars ("SVS" - Soviet Variable Stars).

The present, 4th GCVS edition will consist of 5 volumes. The first three volumes (1985-1987) contain 28435 stars and include information on variables in all 88 constellations. The fourth volume (1990) has just appeared; it is the reference volume for the 4th edition. It contains the list of variables in the order of their right ascensions, the list of variables arranged by their variability types, as well as cross-reference tables enabling a user to find whether a star in one of the major star catalogues (HD, BS, BD, CoD, CPD, etc.) is also listed in the GCVS or the NSV catalogue.

Starting with the NSV catalogue, we are preparing the main tables of our catalogues also on magnetic tapes, and distributing these tapes through the international astronomical data centers. Maybe in future, the remarks and the lists of references will also be available on tape.

We hope to publish soon the fifth (and the last) volume of the 4th edition, devoted to extragalactic variable stars. The prospects of this volume will be discussed at this meeting in the talk by Dr. N. Lipunova.

I would like to emphasize that the preparation of the GCVS and the NSV is not only compilation but critical evaluation of all existing data. This can be illustrated quite well, for instance, by a rather funny case of V600 Aql, "Nova" Aql 1946, which was proven to be an asteroid during the preparation of the present GCVS edition through the comparison of the published photograph with an unpublished one taken an hour later (note that the ephemerides of minor planets for 1946 were never published).

The number of catalogued variable stars increased exponentially for some time; now it seems to increase not so quickly, and I really hope this tendency will turn out to be real.

The GCVS data base has been created over many decades, and historically it became a card catalogue, each card containing information (not only references or annotations!) on one star from one publication. At present we have about half a million handwritten cards, and the volume of information on these cards is estimated as about 0.5 GByte. Of course even an attempt simply to enter all these data into a computer will take much effort. Nevertheless we are really going to undertake such a work, starting with the new information and proceeding into the past, and though we have certain difficulties, mainly in getting good hardware, we hope to provide the users with the data from our complete data base some day.

To finish, I would like to mention that, starting with the 67th Name-list of Variable Stars, we present in our name-lists not only identifications but something like a "mini-GCVS", with data on positions, brightnesses, variability types. We hope to send the name-list No.70 to the publishers in the very near future.

Discussion

Lampens: Does the 1987 version of the GCVS contain the cross-reference tables (which are mentioned in your talk)?

Samus: Presumably you mean the magnetic tape version. Well, the answer is "not yet". Cross-reference tables are in Vol. 4 (1990), and they will be available on tape from Strasbourg rather soon.

Isles: Will the namelists of new variable stars also be available in machine-readable form?

Samus: In principle it is easy for us to make them available; I think we shall do it.

Lowder: What are the plans for any supplements to the GCVS?

Samus: We are still not sure what to do after we have Vol. 5 of the 4th edition published: to prepare a supplement or to start the 5th edition.

Mahmoud: I think that the definition of a variable star is "one which has changed in brightness in the optical region". Because if we use all the wavelength regions, the Sun should be a variable too.

Samus: To be called variable, the star should vary in the optical, UV or IR. The Sun is surely a variable (under any definition).

THE EXTRAGALACTIC VARIABLES IN THE FOURTH EDITION OF THE GCVS

N. A. Lipunova
Sternberg Astronomical Institute
Universitetskij Prosp. 13
Moscow University
119 899 Moscow, USSR

The data on the extragalactic variables will be compiled in the 5th volume of the present 4th edition of the GCVS. As a matter of fact, it will be the first systematic catalogue of extragalactic variables of its type. It will appear, presumably, in 1991. This volume will contain the data on variables in the nearby galaxies, namely the Large and Small Magellanic Clouds, the Andromeda Nebula (M31), the Triangulum Nebula (M33), the dwarf galaxies Draco, Ursa Minor, Sculptor, etc. It was supposed initially to have a deadline of January 1990. In view of forthcoming information, however, particularly for additional galaxies, the deadline will be extended.

In the process of the catalogue's compilation, a great deal of work has been done on identifications of variables and determinations of their co-ordinates. For the LMC (mainly, in its central part containing about 70 per cent of the whole volume), the co-ordinates have been determined by Dr. V. P. Goranski who used the photographic plate obtained in Chile by Pulkovo astronomers before 1973. There is a photographic plate available for the UMi galaxy as well. In other cases, an astrometric technique has been applied for the determination of co-ordinates using reference stars on the basis of published images. Now a plate has become available for us with the image of the SMC, and the determinations will be repeated for this galaxy. Most of the co-ordinates have been verified by Dr. N. N. Samus by using the Turner quadratic method with reference to objects with good positional data. The co-ordinates for variables in the Sculptor galaxy have been reduced from S.L.Th. J. van Agt [DDO Publ. **3**, 205, 1978] by the late Prof. P. N. Kholopov and verified by N. N. Samus.

As expected, the majority of extragalactic variables belong to Cepheid types. Among all stars with more or less reliably defined types, those classified as δ Cep variables constitute about 60 per cent in the LMC, 78 per cent in the SMC, 45 per cent in M 31, and 54 per cent in M 33. The type distributions of the variables in these galaxies are shown in Figs. 1-3. The other types are distributed more or less uniformly, with only slight variations, except in the case of M 31. In the latter case, the novae constitute a considerable part of all classified variables, viz., 38 per cent, which is approaching the abundance of the Cepheid types (about 45 per cent). The dotted lines over the type distributions correspond to ambiguous classifications of variables which are to be verified in the future.

56   *The Extragalactic Variables in the Fourth Edition of the GCVS*

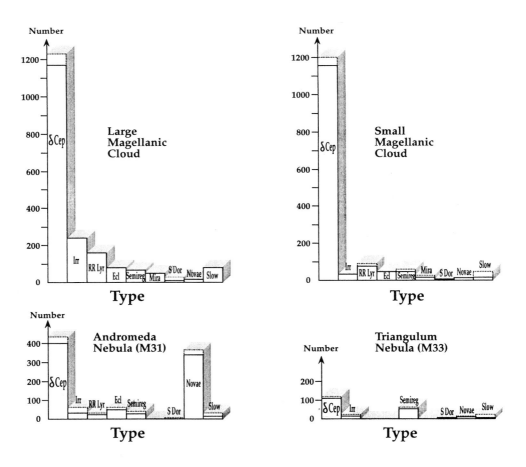

Figures 1 to 3: The numbers of variables of different types in external galaxies.

The classification of extragalactic variables repeats in general lines those accepted in the previous volumes of the 4th edition of the GCVS. It is worth noting, however, the specific new type introduced by us, namely, BL Boo. The stars of this type are encountered, for example, in the SMC often enough. The prototype of the type was discovered by N. E. Kurochkin in the corona of the globular cluster NGC 5466 in 1961 and attributed to the eclipsing binary stars; it was rediscovered by T.I. Gryzunova in 1969. Later, R. Zinn (1974) measured the radial velocity of BL Boo and found that it belongs to the cluster NGC 5466. As a matter of fact, this object is an anomalous Cepheid having too great a luminosity for its period (or too small a period for its luminosity). The variables of this kind are found, in particular, in dwarf spheroidal galaxies. On the period - magnitude diagram for the stars in SMC, BL Boo stars are situated on the left from the main distribution of the δ Cep stars and above the RR Lyr type stars (in the period interval 0^d to 1^d). As to the "Cep" notation, we keep it for the stars which deviate strongly from the main distribution on this period - magnitude diagram.

It seemed to us interesting to look at the period distributions of δ Cep stars as the most abundant class of variables in other galaxies. These distributions for the LMC, SMC, M 31, and M 33 are given in Figs. 4-6. The SMC Cepheids appear to be the most short-periodic, their distribution maximum lying in the period interval 1^d to 2^d. In the LMC the Cepheids are predominating with the periods from 3^d to 4^d, in M 31 with the periods 5^d to 6^d, and in M 33 with the periods 13^d to 14^d. The latter result, however, may not be considered as statistically ensured because of the small number of objects in the group considered.

At the present moment, we have for the extragalactic volume of the GCVS 3595 objects belonging to the LMC, 2319 objects belonging to the SMC, 1176 objects belonging to M 31, 537 variables in M 33, and 604 variables in the Sculptor galaxy. As mentioned above, these were the figures which we had before January 1990. Now we have to introduce additional information concerning some new variables in other galaxies (for instance, about 300 RR Lyr type stars in NGC 185 discovered lately by Hesser and Saha), as well as quite a number of new stars in galaxies already included.

Naturally, the circumstances forced us to restrict ourselves to the minimum set of the most important data, even in the cases of the very thoroughly investigated variables. Partially, we have compensated for this deficiency by giving remarks in the Appendix. For example, we always give information about the interesting behaviour of a light curve, alternative periods in light changes, metallicity, etc.

The main part of the Catalogue will be recorded on magnetic tape which will be deposited in the International Centre of Stellar Data in Strasbourg. The remarks, as before, will not be included in these records and the full information on Catalogue stars is so far available only in our GCVS group.

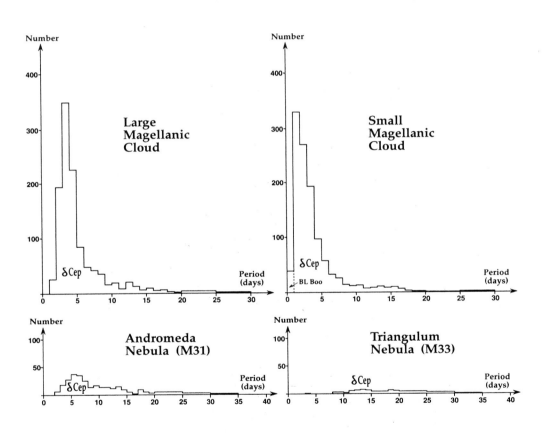

Figures 4 to 6: The distribution of periods of Cepheids in external galaxies.

Discussion

Longo: Are the different period distributions observed in galaxies a consequence of selection effects?

Lipunova: I think that, for the most part, it is due to a real difference between galaxies because of, say, different luminosity functions.

Wolf: The observed period distribution of Cepheids is not only a function of the galactic distance, but also a function of the galactic age. It would be very fruitful to make such statistics for your catalogue of extragalactic variables.

Lipunova: Yes, I think so also. Thank you.

Breger: It needs to be emphasized how important this new volume on variable stars in extragalactic systems will be for the field of variable stars. We are all looking forward eagerly to its publication.

Richter: You spoke about BL Bootis stars. Is this the same as BL Herculis stars?

Samus: BL Her stars are normal short-period Population II Cepheids. They determine the corresponding part of the P-L relation for Population II Cepheids. BL Boo stars are too bright for their periods, even for the continuation of the Population I P-L relation. The two groups are not the same.

Longo: Are supernovae included in the catalogue?

Lipunova: Yes, they are.

Mahmoud: What condition causes the number of variables in galaxies to differ from one another?

Lipunova: I think it is due to a multitude of factors: for example, because of differences in age, morphology, distance, and size, of course.

A CATALOGUE OF VARIABLE COMPONENTS IN VISUAL DOUBLE AND MULTIPLE STARS

P. Lampens
Koninklijke Sterrenwacht van België
Ringlaan 3
1180 Brussels, Belgium

1. Introduction

The compilation of a "Catalogue of Variable Components in Visual Double and Multiple Stars" has been undertaken for various reasons. First, from the practical point of view, such a catalogue of variable and suspected variable stars in visual double and multiple systems is useful for a correct identification of the variable component at the telescope. (In some cases where no absolute accurate positions are available, relative positions - which may be of even higher accuracy - provide helpful information.) This may be especially crucial in performing photometry, not only for a secure identification of the variable component, but also for a good choice of the sky measurement and diaphragm. The experience with the preparation of the Hipparcos Input Catalogue showed how easily even well-known variable stars can be misidentified when they belong to a multiple star system. Indeed, the current variable star designations are not always well cross-referenced to the astrometric catalogue identifiers. (Example: BM Ori in CCDM 05353-0524 (Epoch 2000.))

Secondly, from the astrophysical point of view, we cannot stress enough the importance of studying variable stars that are components of physical double stars. The study of such physical pairs not only allows us to perform differential studies of two stars of equal age, initial chemical composition and distance, but also enables us to reveal some intrinsic parameters of the variable component through the determination of the characteristics of the (less complex) non-variable companion. (Example: a differential study of the variability of two components of the physical pair with colours locating both stars in the instability strip.) In addition, if the pairs are orbital systems, it provides us with information about the masses and thus with some knowledge of stellar structure and evolution, as well as ways to improve the mass-luminosity calibration.

The statistical distribution of variable stars in visual double and multiple stars may also give some clues to the evolution and structure of star systems, as is the case with other statistics on double stars.

To our knowledge, no such catalogue is yet available on magnetic tape. Finally, we have easy access to the astrometric "Catalogue of Components of Double and Multiple Stars", the CCDM (Dommanget 1983; 1989), of which a first version may be expected by the beginning of 1991.

2. Previous Works

The previous works include a "Catalogue of Variable Visual Binary Stars" by Proust, Ochsenbein and Pettersen (1981). The intersection between the "General Catalogue of Variable Stars" (Kukarkin et al. 1969) and the "Index of Visual Double Stars" (Jeffers et al. 1963) made by these authors resulted in a list of 223 stars brighter than magnitude 10. To these, they added fainter stars from the "Catalogue d'Etoiles Doubles ayant une composante variable" by Baize (1962) and stars from Pettersen's (1976) "Catalogue of Flare Stars". The total number of entries presented in their list is 300.

A "Catalogue of Orbital Double Stars with Variable Components" has been recently published by Baize and Petit (1989), based on the intersection of the "Fourth Catalogue of Orbits of Visual Binary Stars" (Worley & Heintz 1983) with the "General Catalogue of Variable Stars" (GCVS) (Kholopov et al. 1985, 1987) and the "New Catalogue of Suspected Variable Stars" (NSV) by the same authors. They provide a list of 199 variable components in 171 orbital systems.

3. The Present Work

The same procedure as used by Proust et al. (1981) has been adopted with the newest catalogues available both for variable stars and visual double and multiple stars. The intersection we made between the GCVS (Kholopov et al. 1985, 1987) and the CCDM (Dommanget 1989) contains more than 400 entries. Accurate positions are provided when available (0.01s in right ascension and 0.1" in declination), as are the component designation and additional identifiers. It has been verified that all stars of Proust et al.'s (1981) catalogue are included.

The remaining work consists of: a) ensuring that all variable components of double and multiple stars in the INCA database (Menessier 1988; Turon 1989) are included. (This database contains about one third of all CCDM systems); (b) including a list of eclipsing variable stars kindly provided by M.-O. Menessier (1989); c) intersecting the NSV catalogue with the CCDM (Dommanget 1989).

Table 1 illustrates the format of the catalogue. The apparent separation and position angle refer to the other component of the system in the case of a double star. Care has been taken when the variable star is the B component to specify the position angle relative to the variable star itself. For a variable component belonging to a multiple system, only one other component is given, chosen in function of the separation and magnitude difference. (For additional information regarding the other components of the system, one should consult the CCDM.) Any astrometric information is extracted from the newest CCDM while the variability information is taken from the GCVS.

Table 1. The Format of the Catalogue

CCDM	HD	GCVS	Position (CCDM)					Variability (GCVS)				DM		ADS
Nr	Nr	Name	Δα ±0.01s	Δδ ±0.1"	θ	ρ ±0.1"	N	Type	Period	Magnitude Sys	To	SP	Nr	Nr
01259+6014A	08538	δ CAS	-506	+76	66°	131.7	2	EA	759.	2.68- 2.76 V	20161	5A5	+59°248	-

4. Problems

One of the problems encountered is the correct identification of the component when no accurate positions exist, i.e. only relative positions are listed in the CCDM. The identification test includes magnitude and spectral type. If no decision can be made due to lack of information, the component column will contain a blank symbol.

Another problem arises when the apparent separation of the components is small. One has to distinguish whether the variability information refers to the global magnitude of the system or to the individual magnitude of only one component. Therefore, the symbols AB, etc... are introduced when apparent separations are too small (in practice, < 6-7" with photometric techniques).

5. Conclusions

Any new information on variability/magnitude/position of a variable component in double and multiple stars is most welcome. We would also like to point out that CCD observations of double stars will be needed in order to identify the variable component in some of the close visual pairs and to study more precisely the light variations of those variable components that present small observed amplitudes, such as the Beta Cephei and Delta Scuti stars.

Acknowledgements

This work is made in collaboration with J.-P. Olivier. We are most grateful to J. Dommanget and O. Nys for providing us with the latest version of the CCDM.

References

Baize, P. 1962, J. Observateurs 45, 117.
Baize, P. & Petit, M. 1989, A&AS, 77, 497-511.
Dommanget, J. 1983, Bull. d'information du Centre de Données Stellaires, Strasbourg, 24, 83-90.
Dommanget, J. 1989, XXth General Assembly of the IAU, Baltimore, Aug. 5, 1988, Eds. A.G. Davis Philip and A.R. Upgren, Contrib. Van Vleck Obs. 8, 77-82.
Jeffers, H.M., van den Bos, W.H.& Greeby, F.M. 1963, Publications of the Lick Observatory, XXI, 2 volumes.
Kholopov, P.N., Samus, N.N., Frolov, M.S., Goransky, V.P., Gorynya, N.A., Karitskaya, E.A., Kazarovets, E.V., Kireeva, M.N., Kukarkina, N.P., Kurochkin, N.E., Medveva, G.I., Pastukhova, E.N., Perova, N.B., Rastorguev, A.S., Shugarov & S.Yu. 1985, 1987, The General Catalogue of Variable Stars, Vols. I-III, 4th Ed., Moscow, "Nauka".
Kukarkin, B.V., Kholopov, P.N., Efremov, Yu.N., Kukarkina, N.P., Kurochkin, N.E., Medveva, G.I., Perova, N.B., Fedorovich, V.P.& Frolov, M.S. 1969, The General Catalogue of Variable Stars, 3rd Ed., Moscow.
Menessier, M.-O. 1988, 1989, private communication.
Pettersen, B.R. 1976, Inst. Theoretical Astrophys., Oslo, Report No. 46.
Proust, D., Ochsenbein, O. & Pettersen, B.R. 1981, A&AS, 44, 179-187.
Turon, C. 1989, private communication.
Worley, C.E. & Heintz, W.D. 1983, U.S. Naval Obs. Publication 24, part 7.

Discussion

Hall: This is a comment. You perhaps should include the SAO number, if your catalogue has space.

Lampens: It would be straightforward to do so, but I am not sure I will because the SAO catalogue is not complete to a given magnitude limit.

Mattei: What is the limiting magnitude of the catalogue and when do you anticipate its completion?

Lampens: The limiting magnitude is observationally defined and depends on the observer. As a general rule, there are no magnitudes below 16. This is also the limit adopted by Proust et al.. We expect a first version of the catalogue to be ready by the beginning of 1991.

CHAPTER 2

VISUAL OBSERVATION

COORDINATION OF VISUAL OBSERVING PROGRAMS

Janet Akyuz Mattei
AAVSO
25 Birch Street
Cambridge, Massachusetts 02138, USA

1. Introduction

The AAVSO was founded in 1911 for the purpose of making the observations of amateur astronomers available to professional astronomers. Although the name is the American Association of Variable Star Observers, the AAVSO is truly an international organization, with members and observers all around the world. Keeping that purpose in mind, the AAVSO, through the decades, has coordinated variable star observations, evaluated, compiled, and processed them, and published and disseminated them to the astronomical community.

The observing program of the AAVSO contains about 3600 stars of all types: pulsating, eruptive, eclipsing binaries, and suspected variables. The stars in the visual observing program have large amplitudes of one magnitude or more, with a limiting magnitude of 16.5. The accuracy of the data is between ±0.2 and ±0.4 magnitude.

The AAVSO receive between 20,000 and 22,000 observations a month, totaling between 240,000 and 265,000 observations a year, from 500 to 600 observers worldwide. The observations are sent to the AAVSO monthly by mail (paper or diskette), electronic mail, or FAX. As the observations come in, they are immediately made computer-readable. At the end of the month, all of the observations received that month are processed and merged with the existing data files on each star and the light curves are kept up-to-date.

To date, the AAVSO has compiled over 6.5 million observations from 1911 to the present, and 4.5 million observations since 1960 are in computer-readable form, residing at the AAVSO Headquarters. The earlier 2 million observations are being computerized.

2. How Visual Observations Are Utilized

The AAVSO receives between 150 and 175 requests from astronomers worldwide each year. These requests are for the following purposes:

1. correlating optical observations with multiwavelength observations obtained from ground-based telescopes or satellites;

2. scheduling observations both for ground-based telescopes and for instruments aboard satellites;

3. providing simultaneous monitoring of observing targets during these scheduled observing runs;

4. analyzing data to determine the long-term behavior of variable stars;

5. obtaining reference materials for science projects or articles on variable stars for books or magazines.

Particularly in the last 15 years, our services in scheduling observing runs, simultaneous optical monitoring of observing targets, and correlation of multiwavelength data have played a vital role in variable star research.

Coordination of observing programs to help astronomers is not a new endeavor of the AAVSO and other variable star observer groups. The AAVSO was doing this during the directorship of Leon Campbell in the 1930's, and also during Margaret Mayall's directorship, particularly in the 1960's with T Tauri stars and long period and cataclysmic variables. The Variable Star Section of the Royal Astronomical Society of New Zealand (VVS, RASNZ), under Frank Bateson's directorship, has been coordinating observing programs, particularly for cataclysmic variables, since the 1950's. Under Ernest Markwick's leadership, the Variable Star Section of the British Astronomical Association (VVS, BAA) coordinated observations as far back as the early 1900's.

The coordination of visual observing programs became particularly important and crucial, however, in the 1970's, when variable star research extended all the way from the x-ray to the radio region of the electromagnetic spectrum. In 1973, a group of x-ray astronomers from Massachusetts Institute of Technology, during a survey of the Cygnus Loop with a sounding rocket, accidentally detected an ultra-soft x-ray source that turned out to be SS Cygni. AAVSO observations indicated that SS Cygni was at maximum at this time of the first detection of x-rays from a cataclysmic variable star. Although theory had predicted that close binary systems like SS Cygni should emit high energies in the x-ray wavelengths, until this experiment there was no observational proof.

That particular experiment started an avalanche of multiwavelength observations of cataclysmic variables, both with ground-based telescopes and with those aboard satellites such as Ariel-5, Netherlands Astronomical Satellite (ANS), High Energy Astronomical Observatories 1 and 2 (HEAO-1, HEAO-2), International Ultraviolet Explorer (IUE), European X-Ray Observatory Satellite (EXOSAT), Voyager, and Ginga. Ariel-5 proved that SS Cygni was a soft X-ray source at quiescence. However, it was not until HEAO-1, which conducted an all-sky survey ranging from the softest x-rays (0.1 keV) to high energy gamma rays (10 MeVs), that eruptive or cataclysmic variables could be more thoroughly studied in the x-ray region. The

HEAO-1 survey indicated that where some dwarf novae were found to be soft x-ray sources, others, like the bright southern cataclysmic variable VW Hydrae, were not.

HEAO-1 was also able to make pointed observations for several hours of observing targets in order to search for their time variability. Some of the unexpected and interesting discoveries from these observations were the 9-second, soft x-ray pulsations from SS Cygni and 32-second ones from U Geminorum, both seen during outburst.

Amateur astronomers worldwide participated in the simultaneous visual coverage of cataclysmic variables, and provided immediate notification of these stars' eruptions, via the AAVSO, to the team of astronomers. Particularly at the predicted times of the outbursts, the observers kept a vigil on them. James Morgan, from Prescott, Arizona, was so determined to catch eruptions of these stars that he set his alarm clock to wake him every hour throughout the night in order to see if one had started. Often, he was the first observer to report the start of an outburst.

The HEAO-2 observing program followed soon after, in 1978. About 200 cataclysmic variables were in the observing program of HEAO-2, which had a much greater x-ray sensitivity and was tightly scheduled. Throughout the mission, the AAVSO was given the schedule of the observing targets. The observers worldwide were in turn notified of these times so that they could provide simultaneous optical observations during the scheduled HEAO-2 observing runs. Observers' efforts were rewarded when correlation between the x-ray and optical observations showed that 70 percent of the stars in the observing program were x-ray sources.

The European x-ray satellite, EXOSAT, an even more sensitive x-ray satellite, was launched after HEAO-2. Again, for this satellite, through coordinated programs, visual observers provided simultaneous coverage for scheduled objects, alerted astronomers to unusual and rare activity in stars observed, and throughout the mission, supplied optical data for correlation.

In 1978, another satellite - the International Ultraviolet Explorer, IUE - was launched to make observations in the ultraviolet wavelengths. In the observing program of this extraordinary satellite, still in operation today, variable stars of all types and amplitudes were given a substantial amount of time. Cataclysmic variables, due to their radiation in shorter wavelengths, were in the principal group of large-amplitude stars observed. The mass loss from these systems and the phase lag in UV emission at the onset of most but not all outbursts of dwarf novae were among the major discoveries made. The discovery of the UV lag has vital implications in the cause of outbursts in these systems.

To check and confirm the presence of the UV lag in the far ultraviolet region, another satellite, Voyager, launched in 1977 to observe solar system objects, was utilized. This satellite was equipped with far-ultraviolet sensors capable of observing stellar objects. Again, observers around the world played a vital role in alerting the interested astronomers to the onset of the outbursts of the observing targets so that the instruments aboard Voyager could be pointed to observe them. Voyager's observations in the far-ultraviolet confirmed that there was a half-day lag between the optical and the far-ultraviolet emission during the observed outbursts of SS Cygni and Z Camelopardalis.

Extensive coordination of observations and very close monitoring of observing targets was also essential to evaluate the short-period oscillations in the optical wavelengths detected only during outbursts of these cataclysmic variables. Periodic and quasi-periodic oscillations with very small amplitudes of only 0.006 to 0.02 magnitude were found. The periodic oscillations had shorter periods ranging from 7 to 39 seconds, and the quasi-periodic longer ones ranging from 23 to 413 seconds. The periods of oscillations varied from star to star and from outburst to outburst of the same star. These oscillations had important implications and were believed to have originated somewhere in the area of the inner disk, and/or boundary layer, and/or the white dwarf.

In 1987, another important discovery in the cataclysmic variable field was made, this time in the radio wavelengths. Coordination of visual observing programs and immediate notification of the behavior and the onset of the outbursts of observing targets enabled the detection of radio emission from EM Cygni during its outburst, thus making it the third dwarf nova, among the 37 observed, for which radio emissions were detected.

Thus, from 1975 to the present, variable star observers around the world, through coordination of observations among them and the AAVSO and the interested astronomers, have played a vital role in the multiwavelength research of cataclysmic variables. They helped in the following outstanding ways:

1. in the first detection of x-ray emission from new members of each cataclysmic variable subclass;

2. in the first detection of soft x-ray pulsations from astrophysical sources, such as dwarf novae;

3. in obtaining the first multi-wavelength spectra of dwarf novae;

4. in the discovery of coherent and quasi-periodic oscillations during outbursts of dwarf novae;

5. in the discovery that there is a delay between the optical and the ultraviolet and far-ultraviolet brightening in some dwarf novae;

6. in the discovery of radio emission from some dwarf novae.

In fact, Dr. Riccardo Giacconi, Project Scientist of HEAO-2, was so impressed with the contributions of amateur astronomers in the observations of cataclysmic variables for HEAO-2 that when he became Director of the Space Telescope Science Institute, he decided to give some of his own discretionary time on the Hubble Space Telescope to amateur astronomers. An Amateur Astronomers' Working Group, consisting of leaders from seven amateur astronomer groups in the United States, was created to review proposals submitted by amateur astronomers. Five proposals were selected and will be given time on the Hubble Space Telescope. I think that this is one of the best recognitions of the contributions of amateur astronomers to astronomy.

A recent observing program that has required one of the most extensive coordination efforts between observers all around the world and the AAVSO is the observation of about 250 long period variable stars with the European satellite HIPPARCOS - HIgh Precision PARallax COllecting Satellite. The mission of this astrometric satellite, launched in August 1989, is to obtain very precise measurements of position, parallax, and proper motion of about 120,000 stars, 250 of which are large amplitude, mostly long period variable stars. To allocate observing time for the targets, the brightness of these objects has to be known in advance. To achieve this, the HIPPARCOS Input Catalogue Variable Star Coordinator, Dr. Marie-Odile Mennessier, initially used 7 years of AFOEV data, and later 20 years of computer-readable AAVSO data and 75 years of AAVSO maxima and minima dates. The predictions were checked against the recent AAVSO observations and discrepancies noted and corrected. HIPPARCOS is now observing these stars each month using these predictions. Every two months the predictions are up-dated and checked against recent AAVSO observations provided by observers worldwide, and any discrepancies between observations and the predictions are corrected. The observation of these stars with the HIPPARCOS satellite is continuing successfully and this ongoing success is, in part, dependent on the continuation of the data from the observers throughout the life of the satellite.

3. Future Coordinated Visual Observing Programs

Coordination of ground-based observations for data support will continue for HIPPARCOS throughout its life (another two or more years) and for IUE for as long as the satellite is in operation.

In addition, when ASTRO-1 is launched aboard the Space Shuttle Columbia, the AAVSO will be coordinating observations of cataclysmic variables, symbiotic stars, and active galaxies, and informing the team of astronomers interested in these stars so that multiwavelength observations may be scheduled during the ten day mission of ASTRO-1. We have been preparing for this coordination since 1986.

This year, another even more sensitive x-ray satellite - Roentgen Satellite (ROSAT) - was launched. From August 1990 to February 1991 ROSAT will be operating in survey mode, scanning the sky for objects emitting in the x-ray region of the electromagnetic spectrum. We have been asked to coordinate observations of cataclysmic variables, which are known to be x-ray objects, that are in the monthly satellite survey fields and to inform the ROSAT astronomers immediately of the onset of their outbursts. These prompt alerts will then enable the astronomers to schedule simultaneous ultraviolet observations with the IUE while ROSAT is observing these objects.

We will also be coordinating observations of cataclysmic variables in support of observing programs with the Extreme Ultraviolet Explorer (EUVE), to be launched in 1992.

Thus, through involvement in coordinated observing programs, amateur visual observers will continue to play an important role in both ground-based and space research on variable stars in this decade.

4. Elements for Success in the Coordination of Observing Programs

The following elements are crucial to the success of coordination of visual observing programs:

1. The most essential element is, of course, the dedicated observers worldwide who can provide 24 hour coverage of the observing targets.

2. Fast communication of data from observers to the association involved, by telephone, electronic mail, or telefax, is necessary for the proper coverage of stellar events.

3. The commitment, expertise, and experience of the coordinator of the programs with the data, with the behavior of the stars, and with the observers is very important.

4. Processing of both past and current observations and the efficient management of the data files are essential on the part of the associations involved, in order to provide quick reference to and prediction of the variable star's behavior.

5. Good communication with professional astronomers well in advance of the observing program is essential. It is important for the coordinator to know, and then to transmit to the observers, the goal and the requirements of the observing run for which coordination is requested.

6. It is essential to have good finder charts with good comparison star sequences for the observing targets. For the success of the coordinated program, and the data collected as a result, the astronomers requesting the coordination should make sure that good finder charts exist in the observing associations. If not, then they should provide them. For example, for the HIPPARCOS mission, Dr.Michel Grenon, HIPPARCOS Input Catalogue Photoelectric Photometry Coordinator,

provided photoelectric comparison star sequences on about 60 stars for which the AAVSO had either poor or no sequences, thus assuring good results in the monitoring of these objects.

7. Funding is essential for the associations that are coordinating observing programs. Staff time, reimbursement of telephone calls, and other expenses incurred have to be covered. We are grateful that several agencies have been funding the AAVSO for this activity. For example, NASA has provided funding for data support either directly to the AAVSO as grants, or indirectly to the AAVSO by me being a co-investigator of various observing programs.

The successful coordination of observing programs, thus enabling observers to make important contributions to the field of variable star astronomy, has been and will continue to be one of the highest goals of the AAVSO. It is particularly rewarding to know that in this highly technological age, amateur astronomers with their small and moderately sized telescopes continue to make important and often crucial contributions to the multiwavelength observations of variable stars.

My professional colleagues, remember that when you acknowledge the AAVSO, RASNZ, BAA, or any other organization, you are acknowledging not only the organization, but also the many silent contributors - the unsung heroes - to your research programs. You are acknowledging many hours of dedication at the telescopes, fighting cold weather, 8 3 mosquitoes, or other undesirable conditions, in order to help you. You are acknowledging the spouses and families that allow loved ones to spend the nights outside observing for you, instead of inside with them. You are acknowledging the true lovers of stars - the amateur astronomers!

Questions and Comments:

Blaauw: With reference to the first part of your talk: At which stage are the visual observations submitted by observers directly for publication in the publications of the CDS [Centre des Données Stellaires] incorporated into the AAVSO Data Bank?

Mattei: The observations incorporated in the AAVSO Data Bank are only those submitted to the AAVSO by the observer. The AAVSO does not place these observations in any Data Bank such as the CDS for these observations have not been checked [evaluated] for accuracy and/or published.

Baldwin: When observers submit observations electronically are they requested to submit two separately keyed copies of the observations so that they can be verified by the AAVSO?

Mattei: No. When observers submit digitized observations, either electronically or on floppy disks, we at AAVSO Headquarters assume that the observer has already verified the accuracy of these observations before submitting them. Even if we were to request two separately keyed copies of the observations, since we do not have the original record of the observations at the AAVSO, we could not decide which of two discrepant observations was correct.

Williams: Are observations sent on a floppy disc counted as electronic receipts or as [mail] receipts?

Mattei: Observations submitted on a floppy disc are considered as an electronic receipt. Since we started to distribute our standardized data-reporting computer program more and more observations are being sent on floppy discs.

ON THE HOMOGENEITY OF VISUAL PHOTOMETRY

C. Sterken
Astrophysical Institute
University of Brussels (VUB)
Pleinlaan 2,
1050 Brussels, Belgium

J. Manfroid
Institut d'Astrophysique
Université de Liège
Avenue de Cointe
4000 Liège, Belgium

1. Introduction

Visual photometry is the oldest form of photometry, and is the most readily available form of photometry available to amateur astronomers. It can be performed with the naked eye, and with any kind of telescope (even with opera glasses). Pritchard (1881) called it ".. an important and hitherto obscure branch of astronomy". Since then, visual photometry has been intensively applied to many branches of astronomical photometry. The flow of visual measurements sent to variable star associations or sections of astronomical associations is still formidable. AAVSO receives about 250,000 measurements per year.

We are now in an era where visual photometry is under question. The availability nowadays of very accurate detectors (photomultipliers, pin-diodes, CCD's) adds an enormous amount of data to those collected by the eyes of thousands of observers, and more and more amateurs prefer these modern electronic detectors instead of their own eyes. But it is well known that when long series of measurements are interrupted in order to make place for more modern equipment, all kinds of calibrations need to be repeated, and finally, the new data may prove to be somewhat incompatible with the older ones (see for example fig. 2 and 3 in the paper on EW Scuti at this conference).

The greatest asset of existing visual data is that they have been collected over a very long time baseline (in principle over thousands of years). But one must be careful: old visual data are not directly comparable to more recent estimates, even though the detector (the human eye) has not changed since the beginning of recording of stellar magnitudes. There are two basic reasons for which modern visual estimates are not comparable to ancient ones. First of all the ancient astronomers (like Ptolemy, Al-Sûfi, Brahe) used discrete classes, and the associated error bars are of the order of 0.5 to 1 magnitude. Second, the definition of the magnitude itself has changed. All famous catalogues until quite recent times used magnitude classes based only on the psychophysical perception of equal brightness steps. Herschel has shown (see Young 1990), that this more or less corresponds to magnitudes being inversely proportional to the square root of stellar brightness. Nowadays, since the work of Steinheil and Pogson, the magnitude is defined as being proportional to the logarithm of brightness. If two stars differ in brightness by a factor of 100, then they are separated by 5 magnitudes, and a difference of one magnitude corresponds to the fifth root of 100, i.e. about 2.52. In the older scale, this ratio was 3.5 between a first and a second magnitude star, and it decreased to 1.4 between a fifth and a sixth magnitude star. Hence, comparison of visual data must begin only from the time that this modern visual magnitude scale was defined.

The world's largest collection of modern visual estimates is at AAVSO: a fabulous data bank with millions of variable star observations supplied over the last 80 years by thousands of observers. This vast amount of historically valuable data makes us dream about the potential scientific use, especially now that those data are being recorded electronically and will soon be available in computer readable form. The question is: "How homogeneous can such a catalogue be?". Homogeneity is very important: scientists using visual data have noticed strong discrepancies in the results listed in tables of visual data. Those discrepancies differ from observer to observer, but also for the same observer anomalies from one epoch to the other do occur. Our limited experience with amateur observations clearly shows personal equations. In the study of the peculiar variable V348 Sgr, for example, part of the scatter in the visual estimates of V348 Sgr presented in Heck et al. (1985) is due to systematic differences between observers (note that the problem related to EW Scuti (see above) has nothing to do with personal error: it is simply a problem caused by different spatial resolution). To an experienced photometrist, used to measuring with highly precise instruments and well-calibrated detectors, the transition from dream to nightmare is only a small step, especially when relying on visual data coming from catalogues of non-homogeneous character. Inhomogeneities in modern visual databanks are caused by errors of three kinds: "detector" errors, due to the shortcomings of the human eye when used in the photometric mode, errors of observation, caused by inadequate observing techniques, and errors of data reduction coming from improper data reduction procedures. Before we discuss in section 5 these three different kinds of errors, we give a short outline of the basic principles of visual photometry, i.e. the process of vision and the perception of brightness and colors. Section 2 deals with the mechanism of night vision from the physical point of view, sections 3 and 4 deal with estimates of light and color from the experimentalists' viewpoint.

2. The process of vision

An image from the outer world is point for point mapped on the retina, the photosensitive surface at the back of the eye, which converts the light into nerve currents which are transmitted for interpretation to the brain. The mind compensates for chromatic aberration, and fills in the sensation of color. The retina consists of a mosaic with millions of individual biological photodetectors of two different types: rod cells and cone cells. The cones are situated in the fovea (a yellow spot in the central region of the retina) while the rods are situated in the peripheral regions. Cones and rods contain different pigments, which absorb radiation.

The rods are involved in the detection of light at low levels of illumination, and they are even capable of signalling the absorption of one single photon. The high sensitivity of rod vision is accompanied by a loss in color vision. The central fovea is blind in night vision due to the absence of rods. A well-known trick of stellar observers is to detect faint stars by looking at them off center.

Cones function at higher light levels and provide very sharp visual definition and color sensitivity. The retina has three kinds of cones, each of which contains one kind of pigment with specific spectral response; they absorb strongly in the blue, green or red wavelength region at respective wavelengths of 430, 530 and 560 nm. The resulting spectral response is a composite of the individual spectral responses of all (types of) cones. This provides the basis for color vision.

Cones and rods have maximal spectral sensitivity at different wavelength, which causes the well-known Purkinje effect. For example, when looking through a spectroscope to a star producing a weak spectrum, one will see it colorless, with the brightest range around 505 nm. A bright spectrum on the other hand is seen colored; the maximum brightness is around 555 nm, and its total width is much greater than the weak spectrum perceived in the former case. This change in wavelength of maximal sensitivity that occurs in going from night vision (scotopic) to day (photopic) vision is called the Purkinje effect.

The human eye is designed to see light intensities ranging from the midday sunlight to faint starlight (the latter is equivalent to the light of a candle at a distance of about 20 km): the eye's working range is more than a billion to one in light intensity. Neither TV nor photographic film match the ability of the eye to record pictures in such drastically different conditions. The eye is provided with some kind of automatic gain control due to the opening up (or closing down) of the iris and due to switching between cone vision and rod vision. It adjusts to total darkness in a time span which approximately coincides with the duration of the gradual darkening during sunset and twilight.

Dark adaptation also strongly depends on the previous level of exposure of the eyes to light: dark adaptation seems to reach a maximum value after about 20 min in the dark as judged by the observer, but careful psychophysical tests showed that sensitivity may increase out to an hour of dark adaptation. The adaptation to bright light, on the contrary, goes much faster.

Faint stars excite rods, not the cones. During the interval between cone and rod threshold all stars are colorless, and once cone threshold is reached, the light gets a chromatic appearance. At red wavelengths, that interval is very small, so either red light is seen, or no light at all.

Binocular summation gives a slight increase in gain due to the summation from each eye (the same applies to monaural and binaural hearing). The advantage must be regarded as statistical: the probability of detection is increased by the use of both eyes.

The spectral resolution of the eye - the threshold value at which two hues of different colors can be differentiated - is about 5 nanometers for a normal eye. Color and brightness sensations are private experiences, and laboratory experiments confirm that the relation between wavelength and hue is not precisely fixed: it not only varies from person to person, but also varies in time. Extreme environmental conditions provoke extreme reactions: Experiments during manned space missions for example have shown that cosmonauts' eyesight becomes less sharp during the first days of flight, with full adaptation occuring after only 2 to 3 weeks in orbit. Spectral resolution of cosmonaut vision seems to degrade at the red end of the spectrum, whereas they show smaller errors in discrimination at the blue end of the spectrum than do ground-based observers. This effect largely depends on the duration of the mission (Vasyutin & Tishchenko 1989). It also seems that, when luminance is increased, the response of the yellow and blue sensitive cones is stronger relatively to the red, so that the violet part of the spectrum becomes less reddish. Prolonged vision also leads to adaptive effects that produce drastic changes in appearance. It also seems that the region of the retina stimulated has a powerful effect upon chromatic appearance.

The presence of the three different blue (B), green (G) and red (R) sensitivity curves can be seen as if the eye functions as a three color photometric system. The R and G "filter passbands" intersect. The problem is that we do not register the three responses individually, but that the "output" is a single sensation of luminance. Ideally one should be able to match every color sensation with a mixture of three primary colors and by summation of the three luminances. The result of such a color match would yield for every estimate (and for every individual observer) a correction. The result of a color match might be described by an equation, which is called the color equation. Knowledge of the color equation would allow the observer to correct his own estimates to the accepted standard system. But in some cases such correction is absolutely impossible, especially when we deal with observers with abnormal chromatic vision. It is well known that about 8% of the male population suffers from chromatic deficiency (only 0.4% of the females suffer from this defect). A quarter of these cases are called dichromatic color mixers: they need only 2 primary colors to make a color match. Their primaries, unfortunately, are not independent. The remaining 75% of chromatic deficient people are called anomalous trichromats. who have deviating sensitivity curves mostly in the red spectral region. Visual observations performed by people with particular chromatic vision must be regarded as the equivalent of photoelectric observations made with a non-standard filter system (two color, or a non standard three-color). It is very well known that such measurements can never be transformed to the standard system, and such data should therefore not be compared with, or combined with other measurements.

3. The perception of star colors and the color equation

Osthoff (1900,1908) did very careful visual color estimates of thousands of stars during the time span 1885 to 1907. He performed estimates using a 34mm and a 108mm refractor, respectively at enlarging power 18x and 40x. His observing procedure was very careful and meticulous: during the visual estimates he would cover his head and the eyepiece with a black towel, and he would write down the colors in complete darkness. Rarely he would switch on a lantern for inspection of a finding chart. The mean duration of a color estimate was 1.5 to 2 minutes, during which he looked at a star until the color impression became invariable. He would never observe during bright moonlight, nor at occasions where the atmosphere was hazy. He used a scale with 10 color "classes", going from 0 for white, 4 for yellow, to 10 for pure red. With this scale he could classify the brightest stars visible in the telescope. With the naked eye he could see yellow stars up to the second magnitude, and orange stars up to mag 2.5.

The color classification was clearly systematically different from one of his telescopes to the other: color differences up to 1.5 color class for the reddest stars were apparent. He also noted that, when using higher magnifying powers at a same refractor, the whitest stars would be more affected than the yellow stars. Also the presence of moonlight would lead to a systematic color difference in the same sense. Osthoff observed variable stars such as Algol, Mira and chi Cygni in order to estimate the influence of the brightness of the star on its estimated color. A redder color (for a same star) is associated with lower brightness, as is well illustrated in fig. 1 which has been constructed from Osthoff's original data.

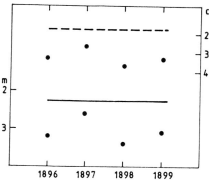

Fig. 1. Variations of color estimate (color class c) associated with light variations (in magnitudes) of Algol. The full line represents Algol's light level outside eclipse, the dashed line represents Algol's color class outside eclipse (based on data from Osthoff (1900).

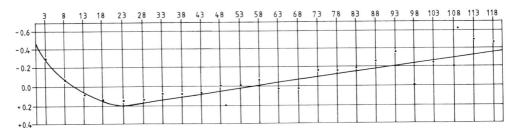

Fig. 2. Evolution of color sensitivity of the eye during one night. The X-axis is the sequential number of the color estimate, the Y-axis gives the color shift (expressed in units of Osthoff's color class).

We have seen that the color response of the eye is not constant: Osthoff's data not only reveal variable response in the course of a same night, but also night-to-night variations, and even variations on a time scale of several years. Very interesting is the evolution of color sensitivity in the course of a night: in fig 2 (from Osthoff 1900) we see how the color of the first stars observed is underestimated by about half a class. The color is correctly estimated after about ten measurements, but a trend to overestimate continues to reach a maximum around the 20th estimate. The individual scatter seems to increase, and on the long run, star colors are registered "paler" than they really are. This effect is most pronounced for orange and red stars, and the amazing point is that the defect is not time-dependent, but only depends on the number of previous estimates. It is clear that we are dealing with minute differences in sensitivity, and they only stand out because of the fact that the eye after hours of dark adaptation becomes extremely sensitive.

Osthoff furthermore noticed that on some nights all color estimates are systematically brighter or fainter than they are on other nights, as if the zeropoint of the eye's color response would change from night to night. Those drifts really manifested themselves as sudden jumps between consecutive nights. A similar effect was also present on a year to-year basis. Fig. 3 (based on data from Osthoff 1908) illustrates this variation of the annual mean drift of the red sensitivity of the eye. The curve has an amplitude of about half a color class, and a characteristic time of about ten years. Surprisingly, seasonal temperature-dependent effects did not appear. But Osthoff did point out that the observer's mental state, physical condition, heart rhythm, blood pressure and stress are important factors in the behaviour of the color sensitivity of the eye.

His extensive data allowed him to compare his results with other catalogues. He gives the following systematic differences between his own catalogue and other catalogues: Dunér 0.83; Krüger 1.3; Schmidt -0.23. Systematic differences of up to one full color class thus exist between color catalogues. To our knowledge,

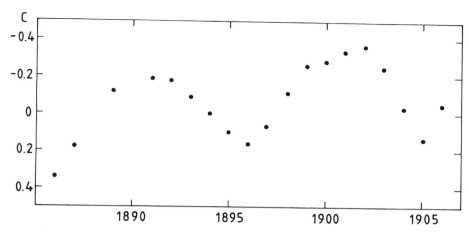

Fig. 3. Variation of the annual mean drift of the red sensitivity of the eye. The Y-axis gives the color shift (expressed in units of Osthoff's color class).

catalogues of color estimates made by female observers do not exist. F.W. Levander (1889) already remarked "It is also a well-established fact that the education of the color sense is much neglected, especially in the case of our own sex". Taking into account the minute occurrence of deviant chromatic vision within female's eyes, it is really regretful that so few female visual observers exist (among the top 25 AAVSO observers - with each more than 33,000 recorded estimates, not a single female observer occurs).

One should not forget that the magnitudes of these systematic differences, and also the individual variations, as discussed above, come from data collected by very experienced observers, and that in the case of poorly trained observers, the effects may be much larger.

4. The perception of star brightness and the visual magnitude scale

The measurement of stellar brightness is based on the assumption that the observer is able to match numbers to his perception, eventually with the help of a suitable standard stimulus. This means that very minute differences between two stars can be expressed numerically. Usually visual estimates of magnitude are only given to one tenth of a magnitude because the eye is not able to detect more subtle differences.

Brightness estimates differ from observer to observer, especially for red objects. As such, each observer has a personal error equation. Pritchard (1884) for example compared magnitudes of 1801 stars measured at Harvard and at Oxford: only 40% agree within 0.1 mag, but 87% agree within 25% or 0.25 mag. The most comprehensive catalogue of visual magnitudes is the great Durchmusterung. But it is generally known that particularly for faint stars, there are considerable deviations from the logarithmic scale. Personal errors, as well as systematic errors introduced by the use of a multitude of instruments with their own optical properties. can be smoothed out by averaging, but one could, by applying proper homogenization techniques, eliminate the systematic errors and obtain average values with reduced mean errors. In such a way one can build catalogues of homogeneous character.

A nice illustration of the systematic differences between observers is the magnitude limit of unaided vision. Older catalogues disagree on these limits. It is generally accepted that the faintest stars visible are of 6th magnitude. Curtis (1901) listed the faintest magnitude in different catalogues on the scale of the Harvard Photometric Durchmusterung. The faintest Almagest star has magnitude 5.38, but Al-Sûfi notes 5.64. More contemporary catalogues score better. Argelander reached 5.74, whereas Heis recorded 1400 stars more than Argelander, and noted 5.84 for the faintest. Houzeau could see stars down to 6.4 magnitude. whereas Gould (at Cordoba) seems to have seen stars of 7th magnitude. Besides the personal condition of the observer's eye, factors such as altitude of observation. cleanness of air and intensity of the diffuse sky background light are important.

It is also easier to see a faint star when you know its direction, and this may explain the reason why Houzeau and Gould could go fainter than the earlier observers. The absolute registered record probably comes from Curtis (1901) who, in an experimental setting, could detect a star of 8th magnitude. Besides Comet Halley at magnitude 19.5, O'Meara saw 20.1 mag stars, with a 24 inch telescope (at an elevation of 13800 feet on Mauna Kea), and from the rest facility at 9000 ft, he consistently saw 8.4 mag stars with the naked eyes. (Sky & Telescope 1985, 69, p.376). These observations rather nicely illustrate the absolute detection threshold of the eye.

The efficiency of light is greater when entering the center of the eye, than when entering along the edge of the pupil. This means that the exit pupil of the telescope and the way the observer places his (her) eye in front of the eyepiece are important and can lead to systematic effects. This effect is called the Stiles-Crawford effect, and is not observed with dim illumination (rods having a wider acceptance angle than cones). This causes an additional systematic effect between large and small telescopes. Every photometric measurement must consist of comparing two light sensations with each other. Absolute necessity is that both light sources match in color and in intensity. When an amateur astronomer performs visual estimates of stellar brightness he only uses his eyes and possibly binoculars or a telescope. He has to tie his measurements to the standard system by comparison with close stars whose magnitudes have been put down on a finding chart. Several techniques exist, well known to amateurs, the most frequently used being bracketing the program star with two stars of similar colors and brightness. Ideally one of those stars would be slightly brighter and the other one slightly fainter. Hence the final value reported by the observer depends on the validity of those comparison stars.

To define the magnitude scale correctly, a zero point is needed. As usual in photometry the magnitude of Vega is defined to be 0.0. Since Vega is not visible from everywhere at any time, secondary standard stars were defined. A convenient group of standard stars, valid for the major part of the Northern hemisphere is constituted by the North Polar Sequence. Stars close to the North Pole are accessible at any epoch of the year. This system of standards, if built from visual observations, is subject to the inaccuracy of the eye estimates. Another system was defined, based on photographic magnitudes. Unsensitized photographic emulsions are very sensitive to the UV and blue regions of the spectrum. On the other hand they do not respond to the yellow and red parts. In order to reproduce the spectral sensitivity of the eye, one has to use photographic emulsions with extended, "panchromatic" response. The overall response is adjusted by using appropriate filters.

Photoelectric photometry is also used to define "visual magnitudes". In fact most photoelectric systems have a bandpass (usually called V) corresponding to the maximum of sensitivity of the eye. This gives another version of the visual magnitudes.

Most of the time, the standard values have not been obtained by visual photometry. They come from photographic plates, or from photoelectric sequences obtained in the particular field. The observer assumes that these magnitudes are close to what would have been obtained by direct comparison with genuine visual standards.

Every telescope will reproduce the colors of the stars slightly differently. This depends on the type of glass or coating, and on the quality of achromatisation. Old objectives often have a yellow- or green-like hue, so systematic effects do occur. These defects are less common nowadays. A great difference of aperture does exert a modifying effect: in a large instrument the color nuances will stand out more clearly. Small scopes will tend to produce more mid-spectrum colors.

5. Causes of errors

All photometry is based upon observation, but not all observation occurs in experimental settings, i.e. experimental conditions are sometimes not adequate, or even not adequately described. It follows from previous discussion that various errors creep into the data. All, or most of these errors affect the homogenity of the data catalogues.

5.1. "Detector" errors

The correlation between visual estimates of brightness and color illustrates that small intensity changes can be recorded where in fact the star is constant. This is clearly illustrated by some observations of long-period variables, where the deviations from the mean lightcurve are much too large to be accounted for by the variable. Whereas the intrinsic variability is confirmed from simultaneous measurements by observers at different sites, this is not the case for the large deviations which have an "instrumental" origin.

The telescope used by the observer acts as a filter with a response depending on the optical elements (coatings, glasses). It can drastically modify the actual photometric system in which the observer is working: refractors and reflectors have distinctly different responses. Also the Stiles-Crawford effect may introduce systematic errors. The atmosphere at the observing site itself is also part of the instrumentation. It plays a major role by reddening significantly the stars. This depends on the altitude, on the state of pollution, sky illumination, etc...To this, one must add the non-constant spectral response of the eye. Different observers have different sensitivity curves, and they vary with time (age, fatigue). Also very important is the response of the organism: depending on age, alertness, motivation and familiarity with the task. Apprehension and excitement at the beginning of the observing run may give way to boredom and fatigue at the end. It frequently happens that an overly cooperative observer reports detection when there is none. Finally. memory or retention may play an important role in the production of accidental errors.

5.2. Observing techniques

The common practice to observe during cloudy nights through "holes" always leads to problems. Thin cirrus clouds are easy to see on a moonlit night, but during a dark night the thin clouds may go unnoticed to the visual observer. Another source of inaccuracies is observing at too large airmasses: the larger the airmass, the more chance that the measurement is flawed by irregular atmospheric transparency. The same of course applies to the situation where comparison stars are too far away from the program star, so that an unnecessary large (and doubtful) extinction correction must be applied. It would be a good practice to observe the stars always at the same airmass in order to change those semi-random errors to systematic ones that could be better corrected. A serious source of error is variable sky brightness which often occurs during rising and setting of the moon.

The use of variable comparison stars of course is fatal. But the magnitudes of comparison stars may be wrong for other reasons: they may be measured in a different system. For instance they have been obtained in Johnson V, or on an inadequate photographic emulsion. Depending on the color of the stars, errors of up to a few tenths of magnitudes can easily occur.

The color of the comparisons may be only poorly known. Ideally, accurate color indices should be available in order to show that they are similar to those of the program star. This information is often obtained by taking only a pair of photographs in blue and in red. This is hardly sufficient considering the inaccuracy of photographic magnitudes. Moreover variable stars often have peculiar spectra and a photographic color index cannot deal with such subtleties.

Also for what concerns the observational procedures and the care with which dark adaptation is looked after, one should realise that the color response of the eye may be totally altered by a preceding session of looking at a TV monitor or by a twilight car trip with headlights on. But not only the eye must be adapted to night conditions, also the telescope must: a necessary, but not sufficient, condition for good seeing with a reflector is that the mirror should not be warmer than the ambient air, so as well the city observer who takes his scope out of a cupboard, as those who travel a distance by car, should let their equipment acclimatize. The presence of bright artificial light on the observing floor may crucially influence the observer's response. Another important effect is present for people who have serious defects in their color vision, like color blindness. It is well known that Weber had an abnormal color response, and this clearly affected his observations.

5.3. Reduction techniques

The differential estimation by using close comparison stars is very crude. It eliminates a few unknowns. such as the first order term of the atmospheric extinction, i.e. the extinction which would affect a monochromatic source at some undefined wavelength in the visual part of the spectrum. That is all. All other effects, mainly due to color terms (in the extinction as well as in the transformation to the standard system), come with full impact. They are more important than in most other photoelectric systems because the width of the passband is so large. Errors grow roughly with the square of that spectral band.

6. Transition era visual estimates / photoelectronic measurement

In stellar photometry we need magnitudes corresponding to definite effective wavelengths, which practically do not vary with the color or the brightness of the stars. The visual system does not satisfy these elementary conditions to the degree that it is fit for standard photometry: the active part of the spectrum is too broad. This is, next to linearity of response, an important reason to turn to photoelectric photometry.

But in the event of using a photometer, the observer must know that if the stars to be compared have different colors, the unaided photomultiplier or dynode - though it is linear and can perfectly repeat its responses, cannot sort out color effects. This does not change when a filter of similar broad bandpass as has the eye is mounted. This is one of the general basic problems of photometry and polarimetry.

A good point to start with is to make homogeneous finding charts, with the same bona fide comparison stars used by everyone. Those magnitudes should conform as much as possible to one "visual" system of narrower bandwidth.

7. Homogenization of different catalogues to a standard catalogue

Müller and Kempf (1907) already stated "measurements from different catalogues should (even if they come from the same observer) not be taken and mixed together without careful study of the relation of the catalogue to a fundamental reference system. The differences between catalogues are not accidental errors, but systematic errors". They stress that every catalogue is to be considered as an independent system and may not be mixed with other catalogues. We would like to add that, whenever catalogues are merged, this only should be done when it is possible to determine the appropriate corrections.

Is there anything one could do to remove or decrease those inhomogeneities? A few remedies are in order, but no complete cure exists. In the comparison of visual results, allowance must be made for the color of the stars, for the color sensitivity of the telescope and its aperture, and for the color perception of the different observers. In determining the corrections for reducing (a partition of) the catalogue to a standard system, attention should be paid to statistical differences, due to the fact that the frequency function of the number of stars common to two catalogues generally varies with magnitude. In practice one should calculate the difference between magnitudes of those stars that have been observed by two different observers (since we deal with variable stars, one should preferably use simultaneous measurements). This number will not be constant, and will even be a function of magnitude: the mean difference between magnitudes estimated by two observers may show a run with magnitude.

Systematic shifts (related to permanent conditions such as the site, or the instrument) can be removed a posteriori. But this is a star-by-star problem, and it is only possible after lengthy series of measurements are available from all parties. In any case, all changes in the equipment (including aluminization, new eyepiece...) should be documented. Pseudo random variations (due to the atmospheric conditions, to large airmass) are almost impossible to deal with. They always lead to changes of the photometric system i.e. to an effective modification of the overall spectral sensitivity curve. This is impossible to correct accurately. Rough correction could, and should, be applied, as in photelectric photometry. but they are quite insufficient.

8. Conclusion

We wish to point out that our emphasizing of precautions should not lead the observer to conclude that all the fun has gone out of visual observing. The overall effect of those errors is large deviations within measurements made at different times by any visual observer, and large, partly systematic, differences between the measurements made by distinct observers.

In that respect the AAVSO files could be analyzed to show the "personal equation" of different observers, and e.g. to look for correlations with the observing sites and/or the equipment. AAVSO database homogenization may teach a lot about competence of individual observers, about the sensitivity of their eyes, and eventually physiological information of ratio of change of brightness with color in eyes with deviant sensitivities. More complete statistical studies based on these extensive files of variable star observations would be very interesting. These goals can only be reached after the catalogue is made available in computer-readable form.

Cecilia Payne-Gaposchkin (1950) wrote: "the existing sequences for variable stars constitute an enormous body of material. Unfortunately, however, very few of these sequences have been adequately investigated as to the system of magnitudes, i.e. as to zeropoint, scale and color equation." We are convinced that conversion of one individual's measurement to the scale of another (or to a standard scale) will maximize the value of each observer's work.

References

Curtis, H.D. 1901, Lick Obs. Bull. 38, 67
Heck, A., Houziaux, L., Manfroid, J., Jones, D.H.P. & Andrews, P.J. 1985. A&AS. Suppl. Ser., 61, 375
Levander, F W. 1889, MNRAS, 33
Müller, G. & Kempf, P. 1907. Pub. Astr. Obs. Potsdam 52.
Osthoff, H. 1900, AN, 153, 3657
Osthoff, H. 1908, AN, 178, 4252
Payne-Gaposchkin, C. 1950, Trans. IAU 7, 267
Pritchard, 1881, MNRAS, 42, 1
Pritchard, 1884. MNRAS, 42, 33
Young, A.T. 1990, Sky and Telescope, March, p 311
Vasyutin, V.V. & Tishchenko, A.A. 1989, Scientific American, July, p66

VISUAL SEARCHING FOR SUPERNOVAE

Rev. Robert Evans
57 Talbot Road
Hazelbrook, NSW 2779
Australia

Introduction

Visual searching provides the best and cheapest opportunities for amateurs to be involved in the business of supernova hunting.

Photographic searching requires expensive and high quality telescopic equipment, and the cost of many films along with their processing and storage. With a Newtonian telescope, not enough galaxies can be observed in a given time to be competitive, although a discovery is always a possibility. The best chances are with a Schmidt camera of ten inches aperture, or greater.

Automatic searching may soon be done by amateurs, but it requires precision engineering in the telescope mounting, a CCD detector, and considerable computer capacity to control all telescope functions, and for galaxy location, image processing and comparison. Usually, several computers are needed at once.

Even when the photographers, and the automatons, have bought and used all their equipment, they still have the same verification problems to overcome as a humble visual observer using a "Dobson" which costs a fraction of the price.

By visual searching, many amateurs with modest equipment will have an opportunity to make a discovery which can be of great value in the world of astronomical research.

The AAVSO is producing its own manual for visual supernova hunting, but the following outline will serve to provide basic details about how to search for supernovae visually, how to verify a suspected discovery, and how to make good use of negative galaxy observations.

How To Conduct a Visual Search

(a) The first necessity is a telescope. A Newtonian of eight inches (f/6) or ten inches (f/5) is the smallest useful instrument. Anything smaller will mean that the observer's chances of finding a supernova will probably be judged as too small to make all of his or her efforts at searching worthwhile, although a discovery will still be a possibility in very close galaxies.

(b) Ready access to a dark sky site is necessary in order to conduct a regular search. Those who can only visit a dark site occasionally will only be able to search occasionally. Those who persist in searching from a light-polluted site deserve a medal. They may be successful occasionally, but will tend to miss the fainter supernovae which might be found from a better site. Obviously, regular use of a dark site is best.

(c) Especially for somebody still learning where the brighter galaxies are located, and even when you think you know where they all are, a good atlas is essential. Tirion's *Sky Atlas 2000* is probably best, at present.

(d) Galaxy catalogues not only tell you where extra galaxies can be found (when you have run out of all the candidates marked on the atlas), but they can also tell you all sorts of little details about each galaxy which may come in very handy during your search routines.

Extra care needs to be taken with those atlases and catalogues which do not tell you the difference between NGC objects which do exist, and those objects which do not exist. There is no point in wasting time looking for a galaxy which isn't there.

Care needs also to be taken with some galaxy magnitudes listed in catalogues. Some magnitude estimates are based upon factors which bear very little relation to what is seen through an eyepiece. So, some galaxies which have bright magnitudes listed for them can be extremely hard to see.

The Revised NGC, Revised Shapley-Ames, the First, Second or Third Reference Catalogue, UGC, Zwicky, Harold Corwin's *Southern Galaxy Catalogue,* or the *ESO Uppsala Survey,* are all useful. The *AAVSO Variable Star Atlas,* and *Uranometria 2000,* are also useful, but, beware of those non-existent objects.

(e) Charts of each galaxy are very important. The key point is that you must have adequate charts and photographs of all the galaxies that you intend to observe. They must be good enough to give you a knowledge of the normal appearance of all your galaxies so that any new supernova can be recognised, and so that all regular features of the galaxy can be seen for what they are. Now that the set of charts by Thompson and Bryan are available, they are a basic necessity.

(f) As well as charts, it is important to have as many good photos of each of the galaxies as you can get. Published photos made with large reflectors cannot always be related to what an amateur can actually see. Schmidt survey photos can have other problems with them, but may often be the best ones available. A photo is especially necessary when no chart of a particular galaxy is available.

(g) It is also necessary to be linked with a team of other observers who have the ability, the willingness and the resources to act as your verification squad. It is no use taking all the trouble to find something if there is nobody to verify it for you. If you leave it till the last minute to arrange these things, your "system" will not work.

(h) Once these requirements are all in place, the basic method of the visual search program is to observe as many galaxies as possible as regularly as possible. Once per lunar month is basic, but twice or three times is better.

To achieve this, speed can be important. So, there are several things which can be done to improve your speed in observing.

A portable, short-focussed telescope is best, which can be used while standing on the ground, without having to move ladders or domes. The mounting should be simple and steady, without slow motion controls or the need to use a drive. Learn to locate objects by star-hopping with a finderscope, and never become dependent upon setting circles. You should learn to be able to locate any galaxy you want in a few seconds only.

After using the finderscope, centre the galaxy in your telescope's field by using a low-powered eyepiece (4x per inch of aperture), and then change to high power, to examine the galaxy carefully (12x to 16x per inch of aperture, or more). Parfocal eyepieces save time by eliminating the need to re-focus.

Learn the normal appearance of all the galaxies "by heart", so that you can instantly recognise anything unusual.

However, the quest for speed should never lead to carelessness, as a faint supernova can easily be missed that way.

Use your resources to check anything about which you are unsure or suspicious.

Verification

All astronomical discoveries are reported through the Central Bureau for Astronomical Telegrams. This Bureau is simply a clearing-house for information, and has no methods of its own to verify any alleged discoveries which are reported to it. So, it is essential to verify thoroughly any discoveries before they are reported to the Bureau.

(a) The discoverer has a number of things to do first. When he or she comes across an object near a galaxy which raises alarm signals in the mind, all of the resources of catalogues, descriptions of objects, charts and photographs, must be used to check whether the suspected object is new or not.

If it is new, it must be checked, as well as possible, in case it is another kind of variable star, or an asteroid. You may know someone with lists of these, if you do not have them yourself. The object should also be watched carefully for an hour in case of any movement.

If, after this, you are convinced that the new object is a supernova, measure from the chart or a photo, the directions and distance of the object from the nucleus of the galaxy. Make a magnitude estimate of the object, and do not pretend that your estimate is any more accurate than can truthfully be claimed. Note the universal time, and the details of the parent galaxy. Then, contact the leader of your verification team.

(b) The verification team leader must make his or her own thorough investigation and observations of the suspected discovery, in order to eliminate the possibility of the suspect being any other kind of object. Other observers should be included in this aspect, if possible.

(c) When it seems clear to the team leader that the suspect is, in fact, a supernova, and that it is not some other kind of object, a report to the AAVSO, and to the Central Bureau, should be drawn up.

The Central Bureau will want to know all the details of all the observations that have been made of the suspect, from first sightings until the verification process was complete.

The Bureau tends to believe most easily those who are already known for their reliable reports. Beginners, especially, must not be surprised if the Bureau asks them for more verification, perhaps even for spectra. If this happens, it becomes important to have some contacts amongst professional astronomers who are both willing and able to obtain these spectra for you. If obtaining spectra is impossible, then getting supporting visual and photographic observations from a number of respected observers is the best way to go.

What Is The Use Of Negative Observations?

Negative galaxy observations are useful as the basis for statistics about how regularly supernovae of the various types appear in the different kinds of galaxies.

(a) The observation schedule has to provide a regular coverage of a large number of galaxies. The regularity has to provide a situation where no supernovae are lost between observations.

Obviously, most galaxies cannot be surveyed for twelve months without a break. But, they can be observed without a break when the constellation is available to be seen. Especially, this can be done by a team of observers who are not all limited by the same weather difficulties.

(b) The number of galaxies observed must be large enough so that a complete range of galaxy types is included. This should include significant numbers of all the major galaxy types. If possible, small galaxies, peculiar galaxies of various kinds, and local dwarves, should be included in the search.

If you want to contribute to an analysis such as this, it is a mistake to search a range of galaxies all of the same type, or which are chosen because they are supposed to produce more supernovae in a shorter time than other galaxies. Such biased samples as this may yield discoveries, but will not be much use as a basis for any useful survey about how often supernovae appear.

An example showing how observations made by amateurs may be used in this way can be seen from two articles which appeared in the Astrophysical Journal. These are:-

Van den Bergh, McClure and Evans. "The Supernova Rate in Shapley-Ames Galaxies." 1987. Ap.J. 323, 44.

Evans, van den Bergh and McClure, "Revised Supernova Rates in Shapley-Ames Galaxies." 1989. Ap.J. 345, 752.

The supernova rate in nearby galaxies might seem a fairly simple matter, but it is poorly known, at present. Even in the Milky Way, the recorded appearance of supernovae over the last two thousand years does not help us much in knowing how often they happen altogether.

The supernova rate is a factor contributing to such things as our knowledge of the composition, structure and motion of the interstellar medium in galactic discs; star formation and their subsequent history; the frequency of pulsars; the expected rate for gravitational wave detectors and, possibly, the origin of heavy elements.

Negative galaxy observations, along with information on time, date, telescope size, and magnitude limit available at the time of observation, should be sent in to the AAVSO.

Acknowledgements

My trip to Brussels to present this paper has been largely financed through the generosity of Continental Airlines, and the Australian Geographic Magazine. This funding has been gained as a result of the persistent help of Mr. Kym Ayling, of Travel Partners - American Express Travel Services, of Penrith, New South Wales.

CHAPTER 3

PHOTOELECTRIC AND CCD OBSERVATION

PHOTOELECTRIC PHOTOMETRY OF VARIABLE STARS

Douglas S. Hall*
Center of Excellence
Tennessee State University
Nashville, Tennessee 37203

During my short 25-minute talk I said what photoelectric photometry is, explained how you can do it and even how much it costs, illustrated how powerful the technique is for studying variable stars, and emphasized that it is an especially excellent way for an amateur astronomer to become a research scientist.

The photoelectric effect is the phenomenon whereby a photon of light strikes the surface of a special material and causes one or more electrons to be ejected. Albert Einstein was awarded the Nobel Prize for his work on the photoelectric effect, not for special or general relativity. If these electrons then flow in a wire, we have an electrical current, which can be measured quantitatively. Thus, photoelectric photometry is using electronic techniques to measure the brightness of a luminous object by means of the photoelectric effect.

Light is not a continuous substance. It is quantized, i.e., composed of individual particles called *photons*. Each photon produces its own burst of electrons, which can be detected. Thus, you have two choices. You can measure the strength of the electrical current. Or you can count the photons. Both techniques are used.

Photoelectric photometry has two major advantages over photometry which uses the photographic emulsion or the human eye as a detector. First, it is capable of greater precision. Precision of 0.01 mag (that's the same as 1 %) is typical and, as we heard during Michel Breger's talk, precision of a few tenths of a percent has been achieved by a few experts (never by me). These experts love to talk of millimag photometry. Second, it is linear. That means the current strength (microamps, for example) or the count rate (photons per second, for example) is strictly proportional to light intensity, over a huge range of intensities. That is not true of the photographic emulsion or the human eye.

There are, however, a few disadvantages. The equipment is more expensive than a camera (and the human eye costs nothing extra). For the same size telescope, its limiting magnitude is not as faint. And, as we heard during Martha Hazen's talk, it cannot study many stars simultaneously, as photography can.

---

* on leave from: Dyer Observatory, Vanderbilt University, Nashville, Tennessee 37235

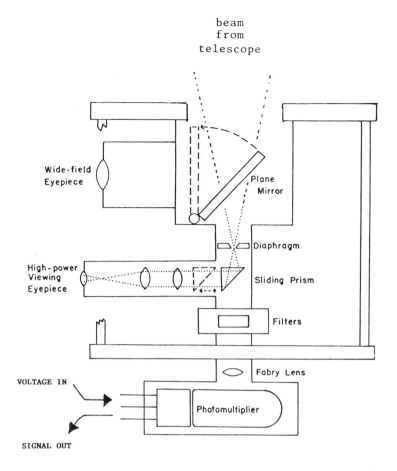

Figure 1. Schematic of a typical photoelectric photometer. The tube which represents the photo-detector is shown, but not the additional supporting electrons.

A complete photometric system (that is <u>everything</u> needed, plus your telescope, to do photoelectric photometry) can be obtained commercially. You can buy it, attach it to your telescope, and begin photometry immediately. Some are even battery-powered and need no external source of electricity. I know of at least five different models, all ready to use, made by four different companies. There may be more. They range in price from somewhat less than $1000 to somewhat less than $2000. References to these are Persha and Sanders (1983), Hopkins (1983), Persha (1989), and Wolpert (1990). Someone skilled in mechanical construction and/or electronics can design and assemble a photometric system, thereby saving money. but that is frankly difficult, time-consuming, and not guaranteed to succeed. Twenty years ago, the only possibility was to design and construct, and the parts were much more expensive and their performance was not as good. So, now is a good time for photoelectric photometry.

Table 1. How to Get a "Delta Mag"

U.T.	target	counts	net	interpol.	Δ mag.
2:03	comp	9004	8953		
---	sky	51			
2:08	var	6191	6138	8832	+0.395
---	sky	53			
2:13	comp	8761	8711		
---	sky	50			
2:18	var	6336	6279	8787	+0.365
---	sky	57			
2:23	comp	8916	8863		
---	sky	53			
2:28	var	6220	6164	8772	+0.383
---	sky	56			
2:33	comp	8734	8681		
---	sky	53			

$$\Delta m = -2.5 \log_{10} \left[\frac{n(\text{var})}{n(\text{comp})} \right]$$

average = +0.381
σ_1 = ± .015
σ_M = ± .009

My first overhead showed a telescope with a photometer attached. The telescope is a Celestron 8-inch (20 cm) and the photometer is an Optec SSP-3, which costs about $850. The small black box contains not only the photodetecting surface but also all of the electronics needed to produce a 4-digit number (on an L.E.D. display) which represents the brightness of your star. This particular telescope and photometer belongs to Martinez Martin, an amateur astronomer in Seville, Spain. Although not reproduced in this paper, the picture can be found in the paper by Martin (1983).

My next overhead, seen in Figure 1, showed schematically what is inside a photoelectric photometer, the black box. The light comes down from the telescope. At the focus it passes through the diaphragm, a small hole 1 or 2 mm in diameter, to isolate the target star from all the other stars in the sky. Then it passes through a filter, so that you can measure light intensity at a known wavelength. And then it hits the photodetector, which converts the photons into an electrical current (which can be measured) or a stream of electrical pulses (which can be counted). In between are simply two eyepieces: first a wide-angle viewing eyepiece, so you can locate your star, and second a high magnification eyepiece, so you can accurately center your star in the small diaphragm hole. When the 45-degree mirrors are flipped away, the light goes straight through to the photodetector. Various other photometer boxes, real ones actually in use, are illustrated in the book by Hall and Genet (1988, chapter 4).

To study variable stars, you get maximum accuracy by measuring the magnitude <u>difference</u> between your variable star and another star (your comparison star) which should be only about 1 degree away in the sky and constant in brightness. This is called *differential photometry*. More details about this technique are given by Hall and Genet (1988, chapter 11).

The overhead seen in Table 1 shows the actual procedure for measuring the magnitude difference between two stars at one time. Most people call this a "delta mag". First you locate your comparison star, center it in the diaphragm, and let the photometer give you a number, which takes about 10 or 20 or 30 seconds. Write this number down, along with the time. Then move the telescope slightly, so that the diaphragm sees only skylight, get its number, and write it down. Next do the variable and its nearby sky, writing down both numbers. Repeat this comp-var-comp-var-comp- sequence, getting the variable maybe 3 or 4 times.

The arithmetic is fun. First, subtract the sky number from the star number, which was actually star <u>and</u> sky. This works because photons or count rates do add and subtract. That gives you the numbers in the fourth column, which are star only. Next, for maximum accuracy, take the average of the two comp star numbers before and after each variable star number, and enter it in the fifth column. The last step is to convert these two numbers in the fourth and fifth columns, representing the light intensity of the variable and the comparison, into magnitude difference. The simple formula is at the bottom of Table 1. In this example we got the delta mag three times. The average of the three is 0.391, meaning the variable is 0.39 magnitudes fainter than the comparison star. The standard deviation of the three tells you that this number, 0.391, is uncertain by about ± 0.009 magnitudes, which is about ± 1%. That is typical for differential photoelectric photometry. Other examples of this procedure can be found in Hall and Genet (1988, chapters 12 and 13).

To do the best possible photoelectric photometry, there are a few corrections which can be applied to this delta mag. One concerns the earth's atmosphere; the other concerns the filter. They are a bit more complicated, but they are usually quite small, not very important, and sometimes can be ignored altogether. For details, see Hall and Genet (1988, chapters 9 through 13).

Table 2. Books on Photoelectric Photometry

Photoelectric Astronomy for Amateurs
F. B. Wood [editor]
(1963) New York: The Macmillan Company

Manual for Astronomical Photoelectric Photometry
A. J. Stokes, D. Engelkemeir, L. Kalish, J. J. Ruiz
(1967) Cambridge: A.A.V.S.O.

Photoelectric Photometry of Variable Stars
D. S. Hall and R. M. Genet
(1982, 1988) Richmond: Willmann-Bell

Astronomical Photometry
A. A. Henden and R. H. Kaitchuck
(1982) New York: Van Nostrand Reinhold

Software for Photometric Photometry
S. Ghedini
(1982) Richmond: Willmann-Bell

Advances in Photoelectric Photometry, Volume I
R. C. Wolpert and R. M. Genet [editors]
(1983) Fairborn: Fairborn Observatory

Micro Computers in Astronomy, Volume I
R. M. Genet [editor]
(1983) Fairborn: Fairborn Observatory

Solar System Photometry Handbook
R. M. Genet [editor]
(1984) Richmond: Willmann-Bell

Advances in Photoelectric Photometry, Volume II
R. C. Wolpert and R. M. Genet [editors]
(1984) Fairborn: Fairborn Observatory

Micro Computers in Astronomy, Volume II
R. M. Genet and K. A. Genet [editors]
(1984) Fairborn: Fairborn Observatory

Lunar Photoelectric Photometry Handbook
J. E. Westfall
(1984) San Francisco: A.L.P.O.

Micro Computer Control of Telescopes
R. M. Genet and M. Trueblood
(1985) Richmond: Willmann-Bell

The Study of Variable Stars Using Small Telescopes
John R. Percy [editor]
(1986) London: University of Cambridge Press

Photoelectric Photometry Handbook, Volume I
D. R. Genet and R. M. Genet [editors]
(1987) Mesa: Fairborn Press

Photoelectric Photometry Handbook, Volume II
D. R. Genet and R. M. Genet [editors]
(1988) Mesa: Fairborn Press

You probably will need a little assistance, to answer some questions, solve a few problems, and achieve maximum accuracy. Also, you will probably need a little guidance or direction, if you want to accomplish serious scientific research rather than just play with your new toy. Both of these are readily available from many sources.

Table 2 lists books on the subject. A few might be out of print and difficult to obtain, but most are readily available and moderately priced. Note that, even 10 years ago, most of these books did not exist. So, you are living at a good time, when help exists.

I also recommend that you make contact with one or more organizations which are involved, at least in part, with photoelectric photometry. There are many, in different countries. In this list is the A.A.V.S.O., which involves itself with photoelectric photometry as well as visual and photographic photometry of variable stars. Many different countries are represented: the United Kingdom, France, Germany, Switzerland, Italy, Spain, the U.S., Canada, Australia, New Zealand, and Japan. Actually, my list here is probably not complete. There may be more. Let me draw your attention to the I.A.P.P.P., which is involved exclusively with photoelectric photometry and has members in 44 different countries, in all seven of the continents. It is a very valuable resource, helping with both "what to do" and "how to do it". At this meeting a poster was presented by Bob Reisenweber, who is Assistant Editor in charge of subscriptions to the I.A.P.P.P. Communications, which is the quarterly publication of the I.A.P.P.P. Membership and subscriptions can be obtained by contacting him (at P.O. Box 8125, Piscataway, New Jersey 08854) or me (at my Dyer Observatory address).

Although the title of my talk was "photoelectric photometry of variable stars", it is exciting and can be scientifically valuable to measure other types of objects as well. This has been discussed by Hall and Genet (1988, table 16-1), but let me list here all those which I know have been observed photoelectrically by amateur astronomers.

Among the variable stars are the eclipsing binaries of the Algol type, beta Lyr type, W UMa type, and zeta Aur type. In the category of rotating variables are (1) the magnetic variables also known as the peculiar A-type stars, (2) binaries varying as a result of the ellipticity effect and/or the reflection effect, and (3) the spotted stars of the RS CVn type, BY Dra type, FK Com type, T Tau type, and solar type. In the category of pulsating variables are the cepheids, the Miras, and the semi-regulars and those of the RR Lyr type, the RV Tau type, and the delta Sct type. In the category of eruptive variables are the novas, the recurrent novas, one supernova (1987A), the symbiotics, the shell stars also known as the B-emission stars, again the T Tau stars, the flare stars, and even the gamma-ray bursters.

Among the non-variable stars are photometric standards and sequences of comparison stars to be used for visual eye estimates of variable stars.

Members of our solar system which have been targeted for photoelectric photometry by amateurs include planets, asteroids, comets, and even the moon.

All sorts of occultation events have been followed photoelectrically as well: occultations of stars by the moon, stars by planets, stars by asteroids, Pluto by Charon, and various moons of Jupiter by each other.

Extragalactic objects have included nearby galaxies such as Andromeda and more distant Seyfert galaxies, which are akin to the even more distant quasars.

Even the sky itself has been a photoelectric target. There have been accurate measurements of atmospheric transparency, some even registering the increased atmospheric extinction caused by high-altitude dust from recent volcanic eruptions. And city-bound amateurs frustrated by very bright skies have obtained scientifically useful quantitative measures of light pollution.

The typical amateur doing photoelectric photometry has a relatively small telescope. The largest I know is 24 inches (60 cm), the smallest is 4 inches (10 cm), and the average is probably about 10 inches (25 cm). With a 10-inch telescope you can easily work with all of the 45,000 stars in the Sky Catalogue 2000.0, still with maximum accuracy. More important than a large aperture is a sturdy telescope mount and a smooth drive. A more complete discussion of the telescope question can be found in Hall and Genet (1988, chapter 3).

Let me finish by giving specific examples of scientific research which has been accomplished by amateur astronomers.

One example, and it is only one example, is the discovery of new variable stars. An overhead shown at the meeting was taken from a recent paper of mine (Hall 1990a, table 1) entitled "Variable Stars Discovered Photoelectrically by Amateurs". As of June 1990, there were 72 new variable stars altogether. For many there is no official variable star designation in the G.C.V.S. because they were discovered only in the last one or two years. More than 40 % have HR numbers and hence are in the Bright Star Catalogue, visible to the naked eye. That fact alone makes them very important stars. The amplitudes of the light variation range from as large as 1.8 mag for one of the eclipsing binaries, to 0.007 magnitude for one of the ellipsoidal variables. Almost 90 % have amplitudes less than 0.3 magnitude, which is the limit for discovery of variable stars by photographic surveys. Photoelectric photometry was necessary to discover these variables. The brightest was magnitude 3.7 in the V, the faintest was 10.9. The shortest period was only 0.1 day and the longest was 4.15 years. In this sample various different physical mechanisms cause the variability: eclipses, starspots, pulsation. the ellipticity effect, and the reflection effect. Some vary by more than one mechanism simultaneously.

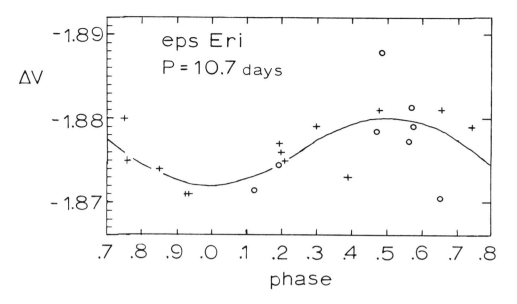

Figure 2. Light curve of epsilon Eridani during the 1989/1990 season. The vertical axis is differential magnitude in the V bandpass, with HR 1098 as the comparison star. The horizontal axis is phase computed with the 10.7-day rotation period. The two different symbols represent the observations of Frey and Wood. The full amplitude is 8 millimags and the rms deviation from the bast fit sinusoid is only 3 millimags.

One area where photoelectric photometry can be very useful scientifically is the timing of eclipses in eclipsing binaries. For a nice introduction, see the paper by Mallama (1987). During my talk I showed the light curve of the eclipsing binary HD 114125, recently discovered by two amateurs in Barcelona, Spain. Though not repeated here, this light curve can be seen in the paper by Casas and Gomez-Forrellad (1989, figure 1). For this work, you determine the time, usually the Julian date, at the moment of minimum light. There are various well-known techniques for doing this as precisely as possible. One method for doing this was discussed by Marvin Baldwin at this meeting.

With many different times of minimum, you can determine the binary's orbital period. This number is useful to know, for example, to predict future eclipses. But the real excitement is that many binary stars have variable orbital periods. Many physical mechanisms can cause this: apsidal motion in an eccentric orbit, orbit of the eclipsing pair around a third star, mass loss in a stellar wind from one of the stars, mass transfer from one star to the other, and radius changes in one star caused by a solar-type magnetic cycle. In a recent paper (Hall 1990b) I reviewed this exciting topic. Let me show two very different examples of variable periods, in two very bright and very famous stars.

At the meeting I showed the O-C curve for beta Lyrae, taken from papers by Klimek and Kreiner (1973, 1975) and by Bahyl, Pikler, and Kreiner (1979). For a nice explanation of how O-C curves work, see the paper by Willson (1986). Each point represents one eclipse timing at one time. The horizontal axis is time, in this case the number of orbital cycles elapsed. The vertical axis is <u>observed</u> time minus time <u>calculated</u> from an arbitrarily assumed constant period. In an O-C curve the <u>slope</u> of the curve indicates the period at any time. In this one the curve is always bending up, so the slope is always increasing, and so the orbital period is always increasing. Since the first eclipse timing, which was by John Goodricke in 1784, until now, the orbital period has increased from 12.89 to 12.94 days, which is 1/3 % longer. The physical mechanism in this case is mass transfer. One star is funneling hot gas over to the other one at a rate of 10 earth masses per year, which is one of the fastest mass transfer rates known for any star. There was a time during the late 1960's and early 1970's when most professional astronomers lost interest in beta Lyrae. Eclipse timings made photoelectrically by amateurs Larry Lovell and Howard Landis were very important in filling this gap.

At the meeting I also showed the O-C curve for Algol, and it is radically different. For a picture of this, see Soderhjelm (1980, figure 2). Again I believe the earliest time is from Goodricke, in 1783. In this case the O-C curve bends down, bends up, down, up, etc. That means the period has decreased, increased, decreased, increased, etc., at the times I marked with arrows on my overhead. Soderhjelm showed that these changes are cyclical, though not exactly periodic, with an approximately 32-year cycle between one decrease and the next, or between one increase and the next. Recent work (Hall 1989, 1990b) suggests that cyclical period change like this result from a solar-type magnetic cycle operating in one of the stars, in this case the cool K-type subgiant. Such cycles are probably analagous to the sun's 11-year cycle. The possible occurrence of solar-type cycles in cataclysmic variables was discussed later at this meeting by Antonio Bianchini. Eclipse timings are extremely valuable in this research area. Because the cycle lengths are rather long, decades, we need eclipsing binaries with a long photometric history. Algol has a 200-year history, and many other eclipsing binaries have 75-year histories.

I had said that not every amateur uses his photoelectric photometer on stars, pointing out that the solar system is also interesting. One of my overheads was a light curve showing an eclipse of Europa by Io, two moons of Jupiter. That work was done in 1985 by the amateur astronomer Frank Mellilo in New York state, with his 8-inch telescope. His beautiful light curve was featured on the cover of the issue of the I.A.P.P.P. Communications which contained his paper (Mellilo 1987). I pointed out that the eclipse duration, from first to fourth contact, was only 10 minutes.

Let me give an example of phenomenal accuracy, which has been achieved by amateurs doing photoelectric photometry. Figure 2 shows a light curve of the single star epsilon Eridani obtained by two amateur astronomers, Gary Frey and

Russell Robb during the 1989/90 observing season. They determined that epsilon is varying with a period of 11.7 days, which is probably the star's rotation period. The physical mechanism is probably a group of dark starspots on one hemisphere, which rotates into and out of view. Note the scale of the vertical axis: each small tick mark is a millimag. The full range of the light variation is 0.008 mag or 8 millimags. And the rms deviation of their points from this curve, which is a sinusoidal fit by least squares, is only 0.003 mag or 3 millimags. This is exceptionally good; I have never seen better; and I have never done this well myself during 25 years of photometry. By the way, Frey and Robb obtained light curves in two other years also. And they observed with two filters: yellow and blue. In all cases they found the same period. In one of the years the amplitude was larger, 0.03 mag or 3 %.

This light curve is not very beautiful, unless you look at the scale. If the photometry has a scatter of about 0.01 mag or 1%, which is larger but more typical, but if the star also has a larger amplitude of variability, then its light curve will look better. I showed such a light curve obtained in 1984 with a blue filter by two other amateur astronomers, Richard Lines and Robert Fried. The star is HD 155638; its variability was discovered by amateur astronomers, and it has recently been named V792 Herculis. Here we have a binary star with an orbital period of 28.5 days. It varies simultaneously by three mechanisms. First: it is an eclipsing binary. The deeper eclipse, primary eclipse, is total. The other eclipse, secondary, is so shallow as to be virtually undetectable, because the cooler star is so much cooler than its hotter companion. Second: it varies outside eclipse in a roughly sinusoidal manner as a large starspot group on one hemisphere of the cooler star rotates with approximately the same 28.5-day period, because the rotation is synchronized with the orbital motion. Third, the light varies with a double sine wave as a result of the ellipticity effect. This happens because tidal forces render one or both of the two stars egg-shaped. This effect is smaller than the other two and difficult to see by eye, but it can be extracted by analysis. The overall amplitude due to all effects is about 0.6 mag. V792 Her is additionally interesting as a particularly intense emitter of x-rays, both hard and soft. They arise from a very hot corona surrounding the cooler star. For a look at this nice light curve, see Bloomer et al. (1983), or Nelson et al. (1990), or the cover of I.A.P.P.P. Communication No. 12.

Many amateurs find it rewarding and satisfying to know they are doing real, useful scientific research, not just playing. Real scientific research is published in a scientific journal, where the results are made available to other scientists. Just to emphasize that amateurs are doing real research with the technique of photoelectric photometry, the I.A.P.P.P. Communications has a regular feature which lists all papers published in astronomical journals having to do with photoelectric photometry and having amateur astronomers as co-authors on those papers. The tally as of June 1990 is shown in Table 3. The first column identifies the publication and the second is the number of papers which have appeared. The total is 289. Let me point out two things. This list underrepresents the scientific contribution by amateur astronomers in photoelectric photometry, because

Table 3. Papers Published in Astronomical Journals by Amateurs

the Astronomer	1
the Observatory	1
die Sterne	1
Acta Astronomica	4
Astronomical Journal	11
Astronomy and Astrophysics	5
Astrophysical Journal	6
Astrophysical Journal Letters	2
Astrophysical Journal Supplement	1
Astrophysical Letters	1
Astrophysics and Space Science	11
B.A.A.S.	9
P.A.S.P.	10
I.B.V.S.	118
Sky and Telescope	1
Telescope Making	1
Byte	1
J.A.A.V.S.O.	30
J.B.A.A.	4
J.R.A.S. Canada	1
J.R.A.S. New Zealand	5
B.A.V. Rundbrief	11
Minor Planet Bulletin	22
Mitt. über Veränd. Sterne	5
I.A.U. Symposia	1
I.A.U. telegrams	1
Cambridge Cool Star Workshops	4
Notes of the A.S.S.A.	1
Monthly Notices of the A.S.S.A.	1
N.A.S.A. (misc.)	4
Annales de Physique	1
G.E.O.S. Circulars	6
Memorie della Soc. Astr. Ital.	1
Nuovo Appunti G.A.W.H.	7

sometimes a project was a cooperative campaign involving so many contributors that all of them could not be included as co-authors, but they contributed nevertheless. Second, notice that there is no such thing as a Journal of Amateur Photoelectric Photometry. Data is data; research is research; science is science. If it is good, it will be published; if it is not good, it won't be published. The good versus bad distinction is not a professional versus amateur distinction.

There are many stars out there waiting for you. There are more stars than astronomers, and always will be.

If you would like to become involved with photoelectric photometry or if you already are and have any questions, please contact me at my Dyer Observatory address. I am waiting, along with the stars.

References

Bahyl, V., Pikler, J., & Kreiner. J.M. 1979, Acta Astr. 29, 393.
Bloomer, R.H., Hanson, W.A., Fried. R.E., Lines, R.D., & Lines, H.C.
    1983, ApJ. 270, L79.
Casas, R. & Gomez-Forrellad, J.M. 1989, I.B.V.S. No. 3337.
Hall, D.S. 1989, in Algols. edited by A.H. Batten (Dordrecht: Kluwer), p.219.
Hall, D.S. 1990a, I.A.P.P.P. Comm. No. 40, 9.
Hall, D.S. 1990b, in Active Close Binaries, by C. Ibanoglu & I. Yavuz
    (Dordrecht: Kluwer). in press.
Hall, D.S. & Genet, R.M. 1988, Photoelectric Photometry of Variable Stars
    (Richmond: Willmann-Bell).
Hopkins, J.L. 1983, I.A.P.P.P. Comm. No. 13, 19.
Klimek, Z. & Kreiner, J.M. 1973, Acta Astr. 23, 331.
Klimek, Z. & Kreiner, J.M. 1975, Acta Astr. 25, 393.
Mallama, A.D. 1987, I.A.P.P.P. Comm. No. 29, 33.
Martin, M. 1983, I.A.P.P.P. Comm. No. 14, 102.
Mellilo, F.J. 1987, I.A.P.P.P. Comm. No. 30, 15.
Nelson, C.H., Hall, D.S., Fekel, F.C., Fried, R.E., Lines, R.D., &
    Lines, H.C. 1990, Astrophys. Space Sci., in press.
Persha, G. 1989, I.A.P.P.P. Comm. No. 36, 28.
Persha, G. & Sanders, W. 1983, I.A.P.P.P. Comm. No. 12, 40.
Soderhjelm, S. 1980, A&A 89, 100.
Willson, L.A. 1986, in The Study of Variable Stars with Small Telescopes,
    edited by J.R. Percy (Cambridge: Cambridge University Press), p. 219.
Wolpert, R.C. 1990, I.A.P.P.P. Comm. No. 40, 23.

Questions

Hazen: Can you convince me that the sinusoidal fit to the data for epsilon Eri is significantly better than a straight-line fit, i.e., no variation?

Hall: One argument is that three different data sets (from two different observing seasons) were analyzed, each in two bandpasses (yellow and blue). Separate period searches for all six yielded best values which were the same within their uncertainties. The value found, 10.7 days, is very close to an earlier independent determination of the rotation period, the 11.2-day value found in a study of the variable Ca II H and K emission intensity. Looking at it another way, the amplitude in the B bandpass was 0.008 ± 0.002 magnitude. In other words, the sine curve fit you saw is significant at the 4-sigma level.

Sareyan: Following Dr. Breger's talk, we should insist again on the fact that a second comparison star be included in the observing schedule. And it should be a full scale comparison, not a check star, i.e., observed as frequently as the variable and first comparison.

Hall: Let me point out first that the observing sequence I illustrated was just that, an illustration. I was addressing the prototypical amateur in the audience, one considering the switch from visual to photoelectric photometry. My message was that the observing process involved in differential photometry is simple, straightforward, logical, and understandable. Responsible photoelectric photometry should, of course, as you say, involve a second comparison star. Normally I recommend the check star approach rather than the second comparison star approach. The relative merits of the two is an old debate. The first minimizes time spent not observing the variable. The second helps damage control if a comparison star proves to be variable. During most of my years of photometry I have been involved with long-term observing campaigns of variable stars, in which case the same primary comparison star was used, was shown to be constant early in the campaign, and verified by some (but not all) of the campaign participants to have remained constant during each subsequent year.

Mahmoud: One of the important factors in photometric reduction is the calibration of the equipment: amplifier, recorder, etc. In your opinion, is it best to do it every night, every week, or every period of observation? Which gives more accuracy?

Hall: The important point is that some correction factors or calibration coefficients vary with time (nightly or even hourly) whereas others change very little or virtually not at all with time. Here is my experience with some of these. (1) The resistors in the coarse gain steps of old DC amplifiers can be temperature sensitive. Thus they may vary during the night, from a warm evening to a cool dawn, and certainly differ from a hot summer to a cold winter. Your coarse gain step calibration should be repeated frequently, or it should be calibrated as a function of temperature, or the resistors themselves should be thermostated. (2) The principal extinction coefficient (k'), as you know, can differ from night to night. It should be determined nightly, unless differential photometry with a nearby comparison star makes differential extinction effects unimportant. (3) The color-dependent extinction coefficient (k'') surely remains constant, or varies much less than your ability to determine its value. Determine it once and for all as accurately as possible and then forget about it. (4) The transformation coefficient in each bandpass (ϵ_v, ϵ_b, etc.) also should remain remarkably constant from year to year, provided you don't change filters, replace phototubes, wash mirrors, etc. Our ϵ_v at Dyer Observatory changed gradually by only 0.01 over a two-decade interval during which the components of our photometric system were not altered. So I would recommend annual determinations, or the use of a rolling average value, or something like that. (5) The photometric zero point is composed of a number of ugly factors, which can be variable on a variety of time scales, difficult to identify, and impossible to calibrate. Examples would be voltage changes, gain drift, and recorder sensitivity. As you probably know, the best defense against these is employment of differential techniques, not only using a nearby comparison star with your variable, but also effectively interlacing program stars with standard stars during all-sky photometry.

Bianchini: Do you think that photoelectric observations by amateurs can provide useful results by monitoring RS CVn binaries to search for solar-type cycles?

Hall: Yes, an observational program like that would be ideal for them. Such cycles could manifest themselves in three ways. First, the degree of variability produced by starspots should wax and wane throughout a cycle. Second, the mean brightness should vary throughout a cycle. And third, as I said in my talk, the orbital period in those which are eclipsing binaries should vary in response to a cycle. The variability involved is on the order of tenths of a magnitude, easy for photoelectric photometry. Over 100 such stars are brighter than 8th magnitude, easy for even small telescopes. And the long time scales (years or decades) are impossible for busy professional astronomers to study during short observing runs at major observatories.

ROBOTIC TELESCOPES FOR PHOTOMETRY

John Baruch
Department of Electrical Engineering
University of Bradford
Bradford BD7 1DP

Opportunities for Robots in Astronomy

There are new areas of astronomy which are being opened up by the development of robotic telescopes.

Parallel multi-waveband observing. The space age has opened up the electromagnetic spectrum to astronomy and revealed a universe which is far from being understood. The visible and radio windows in the atmosphere have now been expanded many times over to open windows in the UV, IR, X-ray and more. Objects which shine brightly in these new windows can, often, only be understood by comparing their fluxes to those in other wavebands, particularly the visible. For objects which are variable, especially the many types that are irregularly variable, the observations need to be made concurrently in the different wavebands.

The demand for concurrent visible observations has increased rapidly over the last few years as satellites move from surveying mode to programme mode and radio astronomers increasingly monitor single objects with high resolution telescope arrays like Merlin in the U.K. The required visible observations are invariably simple photometry which could be carried out by automated telescope systems: robotic telescopes.

Long-period and continuous observing. Robotic telescopes will also open up a completely new area of observations in the time domain. Naked eye observers have long played an important role in extensive observing programmes over extended periods of time, particularly in monitoring variables and searching and patrolling the skies for novae and comets. They will continue to do so. Naked eye observations are much quicker than machines, but cannot match the precision of electronic detectors. For high precision photometry, especially for faint objects, it is necessary to use telescopes with electronic detectors such as CCDs. The current time allocation practices for almost all large telescopes limits the observing periods to a few nights for each programme.

It is already known that the study of what are suspected to be chaotically variable objects; e.g. black holes, cataclysmic variables, active galactic nuclei and other condensed matter objects with accretion discs and jets, will be enhanced enormously by continuous observations. Robotic telescopes organised in a global network (Baliunas et al. 1989) will make continuous multi-longitude observing of single objects a conventional mode of operation. The dramatic improvement in our understanding of the Sun with the helioseismology network developed by Professor Isaak (Elsworth et al 1990) of Birmingham University is an example of the rewards of continuous observations.

In addition to those new areas of work, robotic telescopes will provide a significant improvement in the measurement of all types of variables. Robotic telescopes will greatly expand our understanding of the brighter variables, an area of work long neglected by many observatories.

Programmme for a Robotic Telscope

In this review, I use the term robotic telescope for a system which receives observing requests electronically, monitors the observing conditions, opens the dome and schedules itself to match the requests to the conditions, makes the observations, reduces the data and returns the data with quality indices to the source of the request.

Robotic observing which operates without the interactive support of the versatile intelligence of human beings can only tackle the simplest observations with a single instrument e.g. CCD photometry.

The U.K. is proposing to build a 1-metre robotic telescope using a CCD camera as the main instrument. It is estimated that a 1 minute exposure would provide photometry for objects in the field of view to an accuracy better than 3% to a magnitude of V = 18.5. The U.K. astronomical community was asked to suggest projects which they would like to pursue with this robotic telescope. Over 75 different projects were proposed for such a telescope. This exceeded the available time for the first 3 years of operation of such a telescope by a large factor.

The observing programmes that have been suggested for robotic telescopes are mainly covered by the following headings.

- Rapid follow-up of targets of opportunity e.g. comets, outburst of comet nuclei, novae and supernovae, cataclysmic variable outbursts.

- Maximising the opportunity for concurrent observations with satellites and telescopes in other wavebands. (Of all EXOSAT targets, 60% were requests for simultaneous observations in other wavebands; only 25% of these were able to achieve concurrent ground based observations.)

- Maximising the returns on satellite surveys with ground based follow-up e.g. ROSAT follow-up to determine the orbital periods of cataclysmic variables.

- Regular and frequent monitoring of long-term variable objects, e.g. supernovae and classical novae light curves, stellar pulsations, angular momentum loss in contact binaries, active close binary systems, M supergiants and all types of long period variable stars, quasars and active galactic nuclei.

- Surveys of specific types of objects to take the pressure off larger telescopes, e.g. light curves of asteroids, the orbital properties of the inner satellites of the outer planets, photometry of emission line nebulae, pulsations of white dwarfs, blazars and Hubble-Sandage variables.

- Searches for particular types of event e.g. supernovae, flare stars, dwarf novae, search for Oort Cloud comets and cataclysmic variable turn-ons.

- Continuous monitoring of particularly erratic objects, e.g. gamma-burster error boxes and rapidly variable objects which may display chaotic behaviour e.g. cataclysmic variables in outburst or T Tauri stars.

- The search for life by monitoring solar type stars for Jupiter-like transits.

It is clear that there is a considerable programme of valuable astronomical observations that could be undertaken by robotic telescopes.

The Development of Robotic Telescopes

The basis for the development of robotic telescopes is the dramatic reduction in the cost of computing power that has occurred over the past decade. It is now possible to build robots that will continuously perform simple astronomical observations that are now routinely performed by human observers. These robots can relieve professional and amateur astronomers of dull, tedious and repetitive observations.

In the 60's the Royal Observatory Edinburgh made progress in automating the twin telescopes for photometry. Other efforts were made mainly by amateur astronomers. The first systems that are still operating today were started in 1983. The new International Observatory of the Roque Muchachos on the Island of La Palma in the Canaries installed the automated Brorfelde transit circle, renamed the Carlsberg Automatic Transit Circle. This started operations in 1983 measuring the positions of 1000 stars in a typical 12 hour run to an accuracy of 0.2 arcseconds. (Fogh-Olsen & Helmer 1978)

At the same time Boyd, Genet and Hall (Boyd et al. 1985) initiated full nights of automated observing from Phoenix in Arizona. They have since developed their telescopes and there is a small cluster of them operating at the Mount Hopkins Observatory. They are now sold by the AutoScope Corporation (AutoScope). As well as in the U.S., they have been purchased and modified by groups in South Africa and Italy.

Others have automated their telescopes. The Leuschner Observatory 30 inch (Treffers 1985, Kare 1988) has been used for Berkeley Automated Supernovae Search. It searches for magnitude 18 supernovae and made its first discovery in 1986. Some are building their own automated systems including the Automatic Imaging Telescope (AIT) of Richmond and Filippenko which is being built at Berkeley. It is planned to start to operate this 0.8m system in 1991. The European Southern Observatory is developing their own system, as are the University of Bradford U.K. and the Osservatorio Astronomico di Capodimonte in Naples, Italy. Numerous other groups are investigating the development of automated systems.

The essence of a robotic telescope is the ability to schedule itself and respond to its environmental conditions by optimising its observing programme. It is assumed that a robotic telescope will operate a pool scheduling system. Users will make their requests with their own priorities. The management of the system will rank the requests according to a previously agreed system. This could be peer review supplemented by an agreement that targets of opportunity e.g. a new supernova, would have the highest rank, followed by concurrent operations with observations in other wavebands. A long term programme to look at a particular object every hour might override the concurrent observing and so might a request for multi-longitude observing. Such systems of priorities are evolutionary and can only work through a set of rules, agreed by the users, which are regularly updated in the light of experience. The rules adopted would probably be influenced by the funding of the telescope.

The telescope scheduler will have the whole pool of observations to examine. The system proposed at Bradford uses an AI based scheduler (Barrett et al. 1991) restructured to operate during the day when there is lots of computing time to generate a coarse schedule. During the night a fine scheduler tunes up the coarse schedule to match the conditions. This means that the system will select targets according to their rank, their priority and their position in the sky. Generally objects will only be observed close to their meridian transit unless there is a good reason for not doing so such as a programme of concurrent or continuous observing. Observing schedules that are disrupted by poor weather must be rescheduled to ensure that the highest ranking priority objects are observed if they possibly can be whereas lesser targets can be left for another night.

The V = 12 Barrier

There is a natural divide in the problems of robotic observing at about V = 12. Simple robotic telescopes are limited to observing bright (V < 12) stars. The problem is one of confusion. How can the astronomer be confident that the object which the robotic telescope has observed is the one which he/she wanted to be observed? This problem arises when there is more than one object in the field of view of the telescope. The field of view can be defined as the size of the aperture of CCD detector enlarged by the pointing errors of the telescope. For stars brighter than V = 12 it is probable that within a field of dimension 10 arc minutes that there is only one object as bright as 12th magnitude (Allen 1973). For stars fainter than V = 12 there will be other brighter objects in the same field.

As the astronomer moves to fainter objects the problem of confusion rapidly becomes the key problem in the system operation. If the area of uncertainty, the field of view, is 10 arc minutes, a seventeenth magnitude object may have another 200 objects in the same field including one object of 12th magnitude or brighter.

This problem of confusion has a number of implications. It means that precision photometry in the crowded fields of 17th magnitude objects can only be made with area detectors such as a CCD. Two solutions to the problem of confusion have been proposed. One is that every photometric measurement is accompanied by an image of the field with the target star marked, the other is that field pattern descriptors are included for the pattern of stars around the target star as quality indices with every measurement. In this way the location of the target object with respect to its surrounding objects is defined. Pattern recognition can be used to acquire and confirm the target objects but to keep the computing requirements manageable it is necessary to have high precision telescope pointing.

It is possible to get round the requirement for high precision (arcsecond) in the telescope pointing with good communications. For brighter stars where the mean separation is greater than 10 arc minutes it is only necessary to have a telescope that will point with a precision of a few arc minutes. If it is possible to return the CCD frame to the astronomer for reduction, such pointing precision is adequate for the faintest measurements. Such a mode of operation is more in line with service observing than robotic observing, and the image transfer process is not trivial. A CCD image is typically 500 pixels square with 1000 pixel square formats available. A single image of a half or two megabytes would then take about 400 seconds per megabyte to transmit using modems on a telephone line with favourable network conditions. If the astronomer would accept the limitations associated with compression this may be reduced to about 10 seconds.

Alternatively the images could be stored on magnetic media and posted to the astronomer, but this defeats the objective of automation. It is expected that in the next few years the telephone networks in Europe will support data transfers about 40 times faster than the above using ISDN protocols, and so make image transfer a viable option. This still leaves the astronomer with tedious work of the data reduction.

Automated Data Reduction and Community Confidence

The problem of community confidence in machine observations, which incorporate data reduction, can be reduced to:

1. has the robot observed the correct object? and

2. what are the errors of the reduced data?

Until these problems of confidence are resolved, robotic telescopes will remain the preserve of an exclusive club. The route to winning community confidence in robotic telescope observations for all lies in being able to attach sufficient indices to the measurement to enable the astronomers to reassure themselves about the measurement quality. This requires the provision of data which indicates:

- weather conditions
- object profiles
- magnitudes and errors of the standards
- photon count for the object and standards
- field pattern descriptors for the neighbouring objects
- time and system operation parameters

The solution to some of these problems is very different either side of the $V = 12$ barrier. Here I will concentrate on the problems for the fainter observation programmes, although much is still relevant for the bright photometry.

Confidence in the object acquisition would be supported if every observation returned the reduced data and associated parameters along with a highly compressed CCD image marking the target object.

The alternative is to automate and add intelligence to one of a number of CCD image reduction packages that are available. These generate the photometry parameters of the target object. Most of these use VAX or similar sized computers. For robotic telescopes it is desirable to use a less expensive standard than a VAX and most systems are designed around IBM PCs or compatibles. There are now interactive software packages available for IBM compatible PCs that will reduce a CCD image and produce positions for the objects, their profiles and the relative magnitudes of the objects in the field that match these profiles. They will also separate out close companions using the profile generated from solitary stars. A typical such programme is PC Vista (Richmond).

The effective use of automated data reduction packages requires the support of artificial intelligence evaluation routines. Similarly object acquisition field pattern recognition would benefit from the support of artificial intelligence routines, especially in the confused world beyond the V = 12 barrier.

Community Confidence in Robotics

Assured robotic performance requires the incorporation of advanced technologies such as pattern recognition and automated crowded-field photometry. These developments are at the leading edge of technology and their applications need to win the confidence of the astronomical community. They must be evolutionary. They are ideal for collaboration.

If standard protocols and software interfaces can be agreed, these would define the way in which mechanical, electrical and computing systems are linked together. In this way people would have the choice of adopting the solutions of others or going their own way to improve a section of the system. They can start off with their own system provided they build in the hardware and software interfaces to link into the common system in the way that they wish. Such an arrangement will encourage discussion, mutual assistance and competition within a framework of collaboration. This process will encourage a robotic telescope club collaborating and competing in the most advanced areas of technology.

Robotic telescopes are simple mechanically. Their complexity lies in the sophisticated technology they require. These technologies include networking, AI for scheduling and pattern recognition, tele-operation, image processing, data fusion and robotics. There is considerable industrial interest in all the technologies required for robotic telescopes. They are not expensive to build (about £200K each) but without advanced software to make them autonomous they will remain accessible to only a small group of astronomers.

The development of autonomous robotic telescopes operating in fields of 18th magnitude objects will show that astronomy is not only a culturally and spiritually inspiring subject but it is also of key importance in the development of economically significant new technologies.

References

Allen C.W. 1973, Astrophysical Quantities, 3rd Ed. Athlone Press, London.
AutoScope Corporation, P.O. Box 40488, Mesa, AZ 85274-0488 U.S.A.
Baliunas S.L., Cornell J. & Genet R.M. 1989, "WorldWide Network of Automatic Photoelectric Telescopes", in Automatic Small Telescopes, ed. Hayes D.S. and Genet R.M., Published by Fairborn Press (Phoenix) 1989.

Barrett J.D., Thomas R.C. & Baruch J.E.F., In press. Preprints available from J. Baruch @ Dept. of Electrical Engineering, University of Bradford, BD7 1DP. U.K.

Boyd L.J., Genet R.M. & Hall D.S. 1985, Sky & Telescope, July, p16.

Elsworth Y., Isaak G.R., Jefferies S.M., McLeod C.P., New R., Palle P.L., Regulo C. & Roca Cortes T. 1990, MNRAS, 242, 135.

Fogh-Oslen H.J. & Helmer L. 1978, IAU Colloquium No 48, Institute of Astronomy, University Observatory, Vienna.

Kare J.T., Burns M.S., Crawford F.S., Friedman P.G., Muller R.A., Pennypacker C.R., Perlmutter S., Treffers R. & Williams R. 1988, Rev. Sci. Inst., 59 (7), 1021.

Richmond M. PC Vista, Astronomy Dept. Campbell Hall, Berkeley, CA 94720 U.S.A.

Treffers R.R. 1985, PASP, 97, 446.

MULTICHANNEL PHOTOMETRY FOR AMATEUR ASTRONOMY GROUPS

J.M. Le Contel
Observatoire de la Côte d'Azur
Département A. Fresnel, URA 1361
BP 139 - 06003 Nice Cedex - France

and

E.N. Walker
Deudneys Cottage - Old Road
Herstmonceux - Sussex
BN 27 PU 1944 - England

Introduction

The HR diagram is populated by a great number of groups of variable stars. The contribution of amateurs to their observation is very important in the domain of long period and large amplitude variables. This is related to the fact that the techniques they commonly use are visual and photographic techniques which do not allow very high photometric precision. A consequence of that is a selection in the stellar evolutionary stages which can be studied: only evolved stars and very young eruptive variables have large enough amplitudes to be observed easily by simple observational techniques.

On the contrary many professional astronomers are interested in the first stages of stellar evolution after the main sequence. At this moment of their history, i.e. at the end of hydrogen combustion in the core, many stars become unstable. The determination of the periods allows in some cases the identification of the normal modes of the star and consequently brings information on its internal structure. Two groups are representative of this kind of instability: the β Cephei stars and related B variables (see le Contel and le Contel, this meeting), and the δ Scuti stars. The main characteristics of the variability of these stars are: Periods in the range from 1 hour to 6 hours for β Cephei and δ Scuti stars and from 1 day to 3 days for mid B type variables. The amplitudes never exceed 0.1 magnitude in a B filter. Thus, on a given night no more than one cycle can be observed in a given site. In addition, secular variations of periods and of amplitudes are detected. So long term and longitude-coordinated observations are necessary to get a complete Fourier frequency spectrum and an excellent photometric precision is also needed. The second condition is easy to obtain for professional astronomers while the first is more difficult. It is exactly the reverse for amateurs! How can we improve this situation?

Intercomparison of Results from Different Sites.

It had generally been assumed that if the same comparison stars were used, and the extinction and transformation coefficients were carefully determined, then results obtained

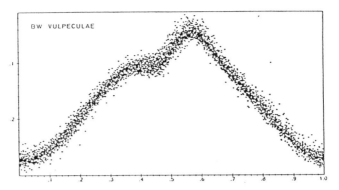

Fig 1 - The light curve of BW Vulpeculae

at different sites, with different photometers, could be put into the same photometric system with adequate accuracy. Regrettably, this seems not to be the case. In a major organisational effort, one of the organisers of this meeting, Chris Sterken, together with 36 co-workers (1986), observed the β Cephei star BW Vulpeculae in 1982 - 1983. Despite a major effort there was a scatter about the mean light curve from all the data of well over $\pm 1\%$ (fig. 1). The amplitudes of the variation in which we are interested for a wide variety of investigations into stellar structure are at, or below, this level. Therefore we are forced to the conclusion that not only is it necessary to obtain extended data runs by multi-longitude observing but also that the equipment used at the various sites must have the same major components. If these two criteria are not met then it seems that the only other way forward is a dedicated photometric satellite or a telescope and photometer based at one of the geographical poles.

In passing it should be noted that for variable stars with very long or very short periods the more sophisticated period analysis methods are not always required. At the top end of the instability strip the Mira or semi-regular stars have periods of tens to hundreds of days and therefore one observation each night is often sufficient to give effectively continuous data. At the bottom of the instability strip the ZZ Ceti stars have periods in the range of, say, two to sixteen minutes. In a six to eight-hour observing run one covers very many cycles, which is a great help in trying to untangle their often complex power spectra. For comparison, in order to obtain the same frequency resolution, a star with a two minute period observed continuously for four hours equates to a star with a period of one hour observed continuously for five days. This would require intense international collaboration.

The Need for a Comparative Photometer.

If we accept that the most rapid way to make progress with our understanding of some of these variable stars is to organise multi-longitude observing campaigns for several days then we have to confront two other problems. One is the fact that very few professional astronomers are able to make themselves available for such campaigns. Secondly photometry is normally only possible in the very best of sky conditions and many of the relatively small telescopes that could be dedicated to the type of observing projects discussed here are not situated in the world's best sites. Indeed, many of the suitable

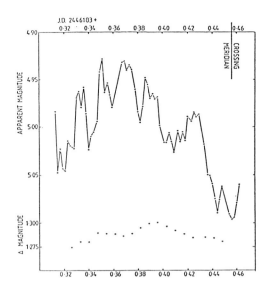

Fig. 2 - Four-star photometer observations of the δ Scuti star 20 CVn (see text).

telescopes are owned by amateur astronomers. We believe, therefore, that the most rapid progress will be made not just by organising week-long observing runs, not just by using the same equipment at the different site, but also by making sure that observers have access to a one-design comparative photometer which will allow observing to continue in poor sky conditions.

Several groups, for example Piccioni et al. (1979), Grauer & Bond (1981), have already built two-channel comparative photometers while Barwig and Schoembs (1987) tested a three-channel one at ESO (Chile). Unfortunately, two channels are not enough to render non-photometric sites photometric. Although it is often clear that the comparison between the two channels shows variations to be present, it is not possible to ascertain whether all the variations are in one channel, or whether both stars are variable. Additionally, for accurate photometry it is not sufficient to have only one comparison star, with perhaps a check star observed infrequently. Instead, unless one reference star can be guaranteed stable to the one millimagnitude level over years, then two comparison stars should be used with equal amounts of time being spent on all three stars. The sky also needs to be observed and we regard four channels as being the correct number for a comparative photometer. Such a photometer is under development in Nice.

The Four Star Comparative Photometer.

For conciseness this photometer is normally referred to as the "Four Star" or 4 ★ and we shall continue that tradition here. The 4 ★ currently under development in Nice, by a team under the guidance of J.M. Le Contel and J.C. Valtier and with assistance of E.N. Walker is based upon the prototype instrument developed by E.N. Walker several years ago. The photometer differs in some major respects from almost all other photometers, the most obvious difference being the use of optical light guides to take the light from the focal plane to the photometer. The professional version of their device has four arms supporting apertures with effective diameters of one minute of arc in the telescope's focal plane. These four arms, plus the intensified CCD TV camera all have XY motions capable of positioning each of the apertures and the centre of the TV frame to less than 20 μ relative to each other. It is only this part of the photometer, plus the ends of the light guides which moves. All other parts of the device, detectors, filters, power supplies, electronics etc... are off the telescope and hence are not subject to varying mechanical loads. Experience using one of the prototype 4 ★ photometers on the 40 cm coude refractor at Nice has shown that comparative photometers of this design are able to produce dramatic improvements in the results with a few millimagnitudes accuracy in very poor sky conditions.

As an example of this we show in Figure 2 the B band results on the δ Scuti variable, 20 CVn, which were taken on a night with cumulus, stratus and cirrus clouds and stars only visible near to the zenith. In the upper part we show the raw light curve from the data with only an extinction coefficient of 1.3, i.e. five times the normal value, removed. In the bottom part of this figure we show the results of the comparative photometry where a newly discovered variation with a period of $\leq$ 70 minutes can be seen superimposed on top of the previously known variation of 2 h 55 m.

The use of light guides bestows another important advantage. There is no need for a Fabry lens in this system, which if the photometer was to be changed to telescopes with different f ratios would require both the changing of the Fabry lens itself and the aperture-lens-detector spacings.

One of the most important features of the 4 ★ system is that after the entrance to the light guides the whole of the system is identical to the JEAP (Joint European Amateur Photometer) which is described elsewhere in these proceedings by E.N. Walker.

Making Comparative Photometers Available to Amateurs.

The 4 ★ photometer is an expensive device and it is unlikely that it could be afforded by any amateur astronomer. The problem is not just the detectors and electronics but the quality of engineering required for the focal plane part of the device is very high. Nevertheless, if amateur astronomers in some of the photometrically underprivileged areas of the world are to contribute to multi-longitude observing campaigns, we have to find some way of providing them with the tools for the job. There are several possibilities which we will discuss below.

1) It might be possible for some professional institutions who have an interest in collaborative multi-longitude observing to buy the photometers and loan them to the amateurs with whom they hope to work. We would recommend that any telescope to be used for this work should have an unvignetted field diameter of at least one degree although the quality of the images at the edges of the field is not important for photometry.

2) If no observatory can find the money to sponsor the project as in 1) above then it might be possible to find either a wealthy corporate or charitable sponsor who could provide the instruments, perhaps on a country-by-country basis.

3) A considerable proportion of the cost of the 4 ★ is the computer-controlled focal plane part of the device. Many amateurs have telescopes that would not be able to take the weight of this assembly, nor would they need the 20 cm diameter coverage of the focal plane that is has. Currently we are trying to discover whether it is possible to build, at low cost, a system of manually controlled apertures which would have the required mechanical stability and cover a diameter of perhaps only 50 mm in the focal plane of a small telescope.

4) There is an interesting fourth possibility which could capitalise upon the similarities between the JEAP and the 4 ★. The 4 ★ system was designed to work on a single telescope with a large field of view. The JEAP was designed to work on axis on telescopes with small fields of view. We note that excluding the number of channels, which necessitates different-sized containers, every component is identical. It is therefore possible to use either a complete 4 ★ system, or a partially populated 4 ★ system less the focal plane assembly, and to equip the telescope ends of the light guides with JEAP telescope interface boxes. (They are described by E.N. Walker elsewhere in these proceedings in greater detail.) It would be entirely possible to assemble two, three or four telescopes, either onto a common mount or mounted separately, and to run a two, three or four ★ system from two, three or four telescopes. A comparison of the costs of four sets of Cassegrain or Newtonian optics, together with a common mount to house them, versus the cost of the full computer controlled focal plane assembly of the 4 ★ suggests that the four telescope option could well be the cheaper.

Note that a two telescope system has been tested by Cl. Gregory of the GEOS group in Haute Provence Observatory (and another has been used successfully for many years by Don Fernie at the David Dunlap Observatory, University of Toronto - *Editor*.

Finally, tests with CCD are presently being made by different groups but the size of the chips is still too small, and rapid CCD are not currently available.

Acknowledgements: J.M. Le Contel acknowledges the organising committee for financial support.

References

Barwig, H., Schoembs, R. 1987,. The Messenger # 48 p.29
Grauer, A.D., Bond, H.E. 1981. PASP 93, 388.
Piccioni, A., Bartolini, C., Guarnieri, A., Giovannelli, F. 1979. Acta Astron., 29, 463.
Sterken, C. et al. 1986, A & AS 66, 11

THE JOINT EUROPEAN AMATEUR PHOTOMETER: AN UPDATE

E.N. Walker
Deudneys Cottage, Old Road
Herstmonceux, Sussex BN27 1PU, England

Introduction

A meeting of the IAPPP for both amateur and professional astronomers was held at Herstmonceux in 1984 (Walker 1985). During this three-day meeting, it became clear that, if the potential for serious research that exists among amateur astronomers was to be realised, then it was necessary to develop a robust photoelectric photometer. The performance criteria which such a photometer must satisfy means that the design is a non-trivial task. In this paper we give a brief description of what has become known as the JEAP (The Joint European Amateur Photometer), and of the design philosophy behind many of its features. (See also Walker 1986).

Design Philosophy

Professional astronomers are finding it increasingly difficult to get observing time on small telescopes for international campaigns, and are turning to amateurs to provide complementary photoelectric observations. Unfortunately most of the amateurs who would like to collaborate with professionals are at a disadvantage because their equipment is not able to produce professional-quality results. The JEAP is intended to have long-term stability and high sensitivity, so that amateurs may contribute to international campaigns and lay down observations which will withstand the test of time. It is also intended to make the JEAP as compatible as possible with a professional photometer. Apart from the acquisition box, which goes on the telescope, the JEAP is identical to the professional four-star comparative photometer described by Le Contel and Walker elsewhere in these proceedings.

Design Criteria

The choice of detector determines the basic features of the design. Professional systems use photon counting for its superior sensitivity, stability and low noise, compared to DC systems. One major contribution that amateur astronomers can make is the laying down of data bases over decades. Some DC systems in careful hands can produce accurate results, but photon counting, which does not have to use variable amplifier gains for different brightness stars, is inherently more stable over long time scales. PIN photodiodes are intrinsically noisy and not a suitable detector for this instrument. Photon counting is therefore used on the JEAP.

The Photomultiplier Tube, Its Housing and Cooling

For UBV work, we have chosen to use the EMI 9924 PMT, or its more modern equivalent, the 9124. Tests on both types of tube have shown that they are capable of giving good single photon discrimination. The more modern tube achieves its best results at about 200 volts below that of the earlier version.

The filters are contained in a black "Delrin" housing which is machined from the solid. This housing is a screwed and sealed fit to the PMT housing which is machined from solid blocks of black anodised aluminium. The base of the PMT, complete with high tension and signal connectors, also mounted on a black anodised aluminium plate, is screwed and sealed to the PMT housing. The reason for the use of these materials is so that the filters themselves are housed in an insulating medium and the PMT is in a housing with good thermal conductivity. This allows cooling of the PMT and temperature stabilisation of the filters. Cooling, with feedback and temperature control, uses thermo-electric Peltier elements and relies on forced air cooling of heat sinks to remove the heat.

The JEAP is designed to house three broad-band UBV or VRI filters as standard, for their higher transmission and greater long-term stability compared to narrow or intermediate-band systems. Other filters may be used but this reduces the compatability of the system.

The photometer must be usable on a wide range of telescope sizes and f-ratios. Additionally, because most amateurs only have access to small telescopes, the photometer must be small and light. A solution to both these problems has been to use fibre optics technology. Experience in developing the four star photometer (Walker 1988) showed that using fibre optics to take the light from the focal plane to the detector was one way to reduce the weight on the telescope. Most photometers use a Fabry lens between the focal plane and the detector. The function of the Fabry lens is to nullify the effect of image movement and spread the image on the detector, but to work correctly it has to be matched to the f-ratio of the telescope. To overcome this restriction, the JEAP does not use a Fabry lens; instead the out-of-focus image is projected directly onto the end of the light guide, which performs the function of the Fabry lens. In this way, the JEAP may be used with telescopes from F/3 to F/~∞.

The light guide is interfaced to the telescope by a small acquisition box. On the front, it has a 1.25" outside diameter tube to fit into a standard eyepiece holder, while on top it has a 1.25" inside diameter eyepiece and graticule holder. Inside the box is a flat mirror with three retained positions. These allow the mirror to 1) prevent light entering the eyepiece during an integration, 2) reflect the light through 90° to the graticule for centering the image, and 3) blank off the apertures preventing light from entering the light guide. At the rear of the box is a series of six apertures with diameters from 0.25mm to 1.5mm in steps of 0.25mm. Behind these apertures is an interface to take the light guide which allows the distance between the focal plane and the light guide to be varied.

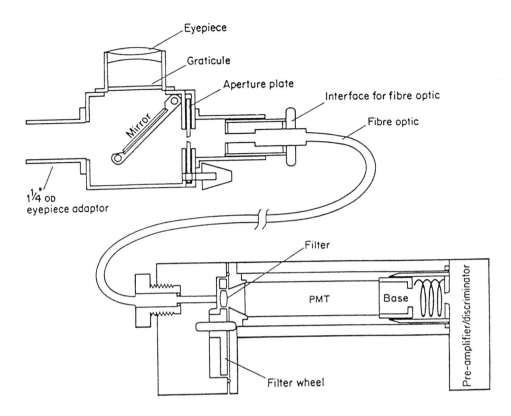

General Mechanical and Electronics Layout

The only part of the photometer which is on the telescope is the small box, described above, and the only connection between this and the rest of the equipment is the liquid light guide. All other parts of the photometer are mounted in a single 19 inch X 3U electronics rack which requires three connections: 1) mains electricity, 2) RS232 serial interface to a personal computer and 3) the light guide. The photometer adds less than one kilogram of additional load to the telescope. Within the electronics rack are the housings for the filter and PMT, with cooling system if provided, as well as all of the electronics.

The electronics consist of a ±5 and ±12 volts power supply, preamplifier/discriminator unit, high tension unit, mother board which carries all timing and serial communication hardware, and a daughter board which carries three eight-bit counters with latches. An important part of the electronics design is that it need only communicate with the control PC occasionally, thus allowing other computer tasks to be carried out.

The high tension units are able to work from either 120/240 AC or 12 volt DC supplies and can be configured to provide either one or two independent HT outputs, each with its own regulation, feedback and LCD output voltage display. Dead time corrections of about 45 nanoseconds are being achieved by the preamplifier/discriminator. The mother board can support up to four counter boards and serves as one of the common components between the JEAP and the four-star photometer. Several preproduction protypes of the JEAP are producing results. Excluding the cooling of the PMT, all mechanical, optical and electronic components are at the production level. The software has worked on a variety of machines, but still is not in a state where it could be sent out to users. Tests of some of the software will be carried out at Nice during the autumn of this year.

References

Walker, E.N., 1985. IAPPP Comm. 19, 20.
Walker, E.N., 1986. JBAA 97, 30.
Walker, E.N., 1988. NASA Conference Publication 100015, 57.

Discussion

Wisniewski. What is the light loss in your fibre optics and how do you guide your telescope since you do not use a field lens?

Walker. The light loss is a function of the length of the light guide and the wavelength. For a two metre long light guide the transmission in the blue is about 90% falling to zero beyond the red limit of the V filter. Guiding the telescope and centering the star are done in the conventional way via the flip mirror and eyepiece on the aquisition box. The telescope should be capable of keeping the stellar image in the middle 10% of the one arc minute, focal plane aperture without guiding for the length of time of the integrations. If this is impossible then the flip mirror could be replaced by a pellicle and guiding would then be possible during the integrations.

Guinan. Can you automate the filter changing and what is the approximate cost of your system?

Walker. Yes, we have provided space and access for automatic filter changing to be included at a later date. The cost depends upon the specification, e.g cooled/uncooled, UBV/VRI etc. In England we are able to assist genuine amateurs likely to produce result with financial help from the Stargazers Trust, a charity set up to help with this kind of work. Once we have the JEAP in real production and if it seems to be of global interest then we will be seeking multinational corporate sponsorship to help with subsidising the price. £3,000 ($5,000) would be about the price of a cooled UBV system with all the software.

EXTRAGALACTIC PHOTOMETRY IN THE TRANSITION ERA BETWEEN PHOTOELECTRIC AND CCD PHOTOMETRY

G. Longo and G. Busarello
Astronomical Observatory of Capodimonte
Via Moiariello 16,
I-80131 Napoli, Italy

and

C. Sterken
Astrophysical Institute
Free University of Brussels
Pleinlaan 2
B-1050 Brussels, Belgium

1. Introduction

In spite of the fact that only twenty years have elapsed since their first applications, Charge Coupled Devices (CCDs, Mackay 1986) have become among the most commonly used detectors for ground-based astronomy. Their success is mainly due to the fact that, at least for some applications, they are within a small factor of being perfect detectors: (i) they have much higher sensitivity than traditional photographic plates (quantum efficiency up to about 80% in the red; a 30-cm telescope equipped with a CCD has about the same light gathering power as a 3 meter class telescope using a photographic plate); (ii) their response is linear over a range of more than 10^5 in input illumination; (iii) they provide digitized data which are easier to handle and to analyze than photographic material; (iv) CCD photometry is more reliable than photographic photometry (Vigroux 1987). Futhermore, recent improvements in the manufacturing technologies as well as the availability of relatively inexpensive image processing facilities, have dramatically reduced the costs of CCD-based observations. This brings CCD technology into the reach of small observatories and even of advanced amateur groups. CCDs have been installed at many small telescopes, upgrading them to unprecedented light-collecting power. We wish to add though that not every small telescope can be rejuvenated as such: upgrading brings with it the tackling of other scientific projects which sometimes require high quality optical and mechanical properties of the telescope, and one must then upgrade only telescopes having those qualities.

While the applications of small telescopes equipped with CCD detectors (hereafter STCCD) to stellar photometry and spectroscopy have been extensively treated in the literature (see for instance, the proceedings edited by Hearnshaw and Cottrell 1986 and Baluteau and D'Odorico 1987), not much attention has been paid to the wide range of possible applications in the field of extragalactic astronomy.

In this review we therefore focus on the following topics: what kind of basic extragalactic research can be performed and what are the minimum requirements for the CCD detectors to be used. Moreover, we present some recipes to obtain accurate and reliable data. More specific problems have been addressed in several other papers (see for example the guide to the specialized literature compiled by Rufener 1989).

2. Some technical aspects

CCD images are frequently affected by gross errors (i.e. bad pixels) due to several factors: cosmic rays (due to either local radioactivity or cosmic events), bad performance of the chip (usually in the form of bad columns or hot spots) and electrical disturbances during the read-out phase. In order to produce consistent scientific data, all these effects need to be kept under control, and the data frames must be corrected appropriately.

The main sources of noise in CCD exposures are the *thermal noise* and the *read-out noise*. All these aspects will be analyzed in more detail in section 4.

2.1 CCD format

Most commonly available CCDs are about 512 by 512 pixels, the linear dimensions of the active image depending on the pixel size, but usually not exceeding 1.5 cm. The typical angular size (out to an isophote of 26.5 mag arcsec^{-2}) of a nearby galaxy is ~ 3 to 4 arcmin, which implies that for focal lengths longer than about 7 m the image will not fit onto a single CCD frame, thus hindering an accurate evaluation of the sky background.

2.2 Pixel size

CCD pixel sizes range from 15 to 30 microns. Atmosphere, telescope, filter and detector modify the distribution of radiant energy. The image of a point source then becomes a very complicated function, the *point-spread-function* (= P.S.F.) which is not known with mathematical exactness. Given a typical seeing of 1 arcsec, the need for oversampling the P.S.F. requires focal lengths between 10 and 15 meters (Walker 1986). If this condition is not met, photometric accuracy falls off rapidly. On the other hand, this condition is not always necessary for the photometry of extended objects and instruments having shorter focal lengths can then be used. It must be kept in mind, however, that oversampling the P.S.F. may affect the determination of colour gradients in the sharply peaked nuclear regions and even the evaluation of some large-scale parameters, like, for instance, the bulge-to-disk ratio in lenticular and spiral galaxies.

2.3 Exposure times

The optimum exposure time is a complex function of several factors, namely: the overall efficiency of the optical system (telescope+filters+CCD), the brightness and colour of the object, seeing, sky brightness and CCD read-out noise.

Usually, extended objects require long exposures and therefore the sky noise dominates over the read-out noise. If this is the case, several exposures of the same field may be combined in order to eliminate cosmic rays, to increase the dynamical range, and to better evaluate the internal errors. It must be pointed out, however, that *the smaller the telescope, the more difficult it is to fulfill the above condition and therefore the use of CCDs with very low read-out noise is more important for small telescopes than it is for large ones.*

2.4 Pixel-to-pixel variations

Even though each CCD pixel has a linear response, there are strong pixel-to-pixel variations in the individual responses, so every pixel must be individually calibrated. To correct for these variations requires dividing the scientific image by a "flat-field" frame, i.e. the image of a uniformly illuminated field taken in the same operating conditions as the scientific one.

2.5 Filters

STCCD are *a priori* confined to broad-band photometry like for instance Johnson UBV and Cousins-Kron RI photometry. The choice of a system of filters matching as closely as possible the standard ones does not need to be over-emphasized: reddening effects are unpredictable, and later conversion of homemade systems to the standard often is simply impossible. When filter-sets deviant from the standard one are being used, best results are obtained if each observer consistently uses his own system. In any case, the colour equation of the system needs to be evaluated by means of comparison to a set of standard stars spanning a large interval of apparent magnitudes and colours.

An important cause of errors is *red leaks*: parasite spectral leaks (very common in glass filters, such as Johnson B) which occur several hundred nanometers redward of the standard passband. In classical photoelectric photometry such leaks are suppressed by using a photomultiplier with low efficiency in the red. Leak signals can be substantial, especially if the spectral energy distribution of the source (galaxy) and the detector response (CCD) peak in the red spectral region. In particular, since the flatfielding is done with the use of a cool lamp, a blue flatfield may be severely biased by the presence of a red leak (unless appropriate colour-balance filters are used to modify the spectral characteristics of the flatfield lamp).

2.6 Linearity and stability

CCD's greatest advantages are the linearity, the stability and the large dynamical range. Minor deviations from linearity are usually observed at low charge levels, due to variations in the charge transfer efficiency, and at very high (half way to saturation) charge levels. In some CCDs, non-linearity at low light levels may be mimicked by delays of the camera shutter. If short exposures are required by the observing program, the shutter dead-time needs to be carefully evaluated.

Stability of CCDs has been tested by several authors: CCDs turn out to be much more stable than traditional photomultipliers (Walker 1986).

3. Some research topics

Even though in the past few years much work has been devoted to analyzing in detail the virtues and vices of almost every aspect of CCD photometry of extended objects (Capacciolo 1989), not enough attention has been paid to the scientific potentials of telescopes below the 1m class, which, in spite of all their drawbacks, may offer four advantages: (i) they are cheaper and therefore more numerous; (ii) they are usually relatively undersubscribed, whereas large telescopes are very difficult to access; they tend to be used to study faint objects only, and time allotments are usually short and very difficult to obtain; (iii) at least in most cases, smaller telescopes have a larger scale in the focal plane and, finally, (iv) they can observe bright objects which would saturate CCDs in the focal plane of larger instruments.

With these points in mind it is easy to identify some types of research which are within the reach of the collecting power of small telescopes and are too time-consuming for large telescopes.

3.1 Imaging programs

Direct imaging is among the simplest applications, requiring only imaging and not necessarily the highest photometric accuracy. Firstly, we would like to put particular emphasis on a research program which has always attracted the attention of amateur astronomers: namely a coordinated search for supernovae by monitoring a large number of selected galaxies or groups of galaxies. After the death of F. Zwicky and until the excitement induced by the anomalous behaviour of SN1987a, systematic surveys for SN had been almost neglected. Nowadays, only a few thousand galaxies are more or less periodically surveyed by both amateur astronomers (Evans 1986, Spratt 1986) and specialized professional groups. The implementation of a few specialized STCCDs would allow improvement in both the number of galaxies observed and the time resolution of the survey, thus leading to better estimates of the supernova rates and of the correlation between supernova rates and morphological types of galaxies. Early detections would also allow the SN to be observed spectroscopically with larger instruments. Furthermore, the use of STCCD would offer several advantages with respect to more traditional techniques:

1) Due to the linearity of the detectors, SN could be detected also very close to the bright central regions of early-type and active galaxies. The detection or the (statistically meaningful) non-detection would be a test for many theories on stellar formation (Longo, Capaccioli and Busarello 1991) and on the powering sources of active nuclei.

2) SN 1987a has proven to be a rare underluminous type II supernova. So far, no other SN with similar characteristics have ever been detected. One possible explanation is the one of a selection effect: most SN are detected visually and the observed difference of 4 mags in the maximum luminosity would make such discovery impossible. In theory, a STCCD larger than 50 ~ 60 cm should allow detection up to the distance of the Virgo cluster.

3.2 Catalogues of global properties

One of the urgent needs of extragalactic astronomy is the necessity of large homogeneous sets of photometric and morphological properties like, for instance, integrated magnitudes, colour indices, isophotal diameters, etc... (de Vaucouleurs et al. 1976). So far, these data have been obtained either from photographic material or from multiaperture photo electric photometry (Longo and de Vaucouleurs 1983, 1985). In most cases the derived quantities (especially the diameters and the total magnitudes) are strongly affected by systematic errors as well as by the non-homogeneity of the raw data which, in many cases, are barely better than the visual (eye) estimates. The availability of large amounts of STCCD's observing time makes it possible to overcome these problems and to obtain quantitative measurements in an homogeneous system (Djorgovski 1986). Furthermore, the determination of several parameters, like for instance the total magnitudes, requires the careful evaluation of the sky level and, therefore is possible only if the field of view is large enough to allow a reliable determination and subtration of the sky background (Capaccioli 1989) - a condition that for nearby galaxies is rarely met by large telescopes. Equally useful are statistical studies of large number of galaxies for peculiar morphological features, like, for instance, dust lanes in early-type galaxies, tidal distortions, peculiar nuclei and so on.

3.3 Morphological and geometrical parameters

As already pointed out by Djorgovski (1986), STCCDs may prove quite useful in obtaining for a large sample of galaxies much needed specific data like, for instance: (for early-type) degree of central concentration, anomalous distortions, etc - all kinds of data that, due to the saturation of the images recorded in most photographic surveys cannot be derived from large scale Schmidt plates and need to be measured on carefully taken images of the single galaxies.

4. Some basic recipes to obtain good CCD data

4.1 Data reduction

As mentioned above, the scientific use of CCD data requires some non-trivial preprocessing consisting of several steps:

1) Subtraction of dark current and bias. The first concept is a signal proportional to the thermal electronic noise: dark current is proportional to the exposure time. The second is an artificial (in a first approximation, constant) charge level added to the pixels in order to extract the signal. Biases may be evaluated by taking very short exposures (1 sec) with the camara shutter closed. Usually bias correction does not require a pixel-by-pixel correction but rather the subtraction of a constant value from each recorded charge.

Dark frames are produced in the same way by taking one or several long exposures (of length comparable to the scientific ones). Attention must be paid to avoid the introduction of spurious noise.

2) Flat-fielding implies the division of the scientific image by another image, in order to correct for the non-uniformities of the CCD chip; thus if flat-fields are improperly taken this may lower the signal-to-noise (S/N) ratio. Flat-fielding must be performed for each filter in which observations are made. Several flat-fielding strategies have been tested and each has its pros and cons (Ortolani 1987, Meurs 1987, Grosbol 1987). Correct flatfielding is very important, since the residual flatfielding errors are the limiting factor in CCD photometry. In the case of extended objects, however, the best way is to use the sky itself. This technique consists in taking many frames of a sky region relatively empty of stars, slightly offsetting the telescope between different exposures. By averaging the different frames after having removed the events caused by cosmic particles, all faint stars present in the field will be smoothed out. Due to the perfect matching between the colours of the flat field and of the background in the scientific exposure this technique is also the most suitable to correct for fringing (i.e. for the interference effects produced inside the CCD chip) effects. The main shortfall of this technique, namely the fact that the frames need to be taken at night, therefore consuming useful observing time, is a less severe drawback for STCCD than for larger telescopes.

4.2 Standarization of the results

Intensive imaging work on a long time baseline has no value unless the results are on a uniform scale and on a standard system. In order to produce high accuracy with imaging, one must first of all obtain high accuracy with standards.

Because of the linear response of CCD detectors, standardization poses less problems than in the case of photographic emulsions: whereas the centers of the photographic images of galaxies and of bright stars are saturated, CCD frames (if properly exposed) yield unsaturated pixels. This not only increases the information content of the data frame, it also allows the determination of the stellar magnitude of the inherently bright standard stars. Note, however, that proper standardization is only possible when no red leaks are present in the filter spectral transmission curve.

Standard stars are bright, and bright stars are the ones with the best determined magnitudes. Since these exposures (even on small telescopes) will be short compared to the exposures of the extended objects, one must be sure that the exposure time absolute error is only a couple of milliseconds. If this condition is not met, non-uniformity effects between core and centre of the field will be introduced. The short exposures on standard stars raise another problem when compared with long exposures of other objects: during long exposures the airmass changes over a substantial amount, and this effect is difficult to correct for. One solution is to observe several faint "secondary" standard stars both during short exposures, and during longer exposures (with comparable air mass range as the scientific exposures). This should allow to determine a correction factor. In case that the secondary stars are being taken from a frame with a galaxy exposure, one should make sure that the profile fitted to the stellar image is not contaminated by the underlying outer regions of the galaxy. Such contamination is readily visible as a skew and asymmetric profile. In order to cancel out residual pixel-to-pixel variation, standard stars should be observed several times at different locations on the detector, and also at different airmasses.

Once these basic precautions have been taken into account, one can proceed to place each night's results on a common internal system, of which the zero point can be determined by digital aperture photometry on several stars in each field. This instrumental CCD system can then be transformed to a fundamental standard system. It is obvious that, for a given aperture, apertures for all standard stars should be the same. This standardization can be repeated for a sequence of different apertures, so that for each aperture diameter a valid transformation to the standard system is constructed. This will enable true multiaperture photometry.

Acknowledgements

The authors wish to thank E. Cappellaro for pointing out some possible applications of STCCD to supernovae searches, Tom Oosterloo for clarifying some aspects of the flat-fielding procedure, and M. Rigutti for critically reading the manuscript.

References

Baluteau J.P. & D'Odorico S. (eds) 1987. "The optimization of the use of CCD detectors in astronomy", ESO conference n. 25.

Capaccioli, M. 1988. Boletin de la Academia Nacional de Ciencias, Cordoba, p.317.
de Vaucouleurs, G., de Vaucouleurs, A. & Corwin, G.H. Jr. 1976. "Second Reference Catalogue of Bright Galaxies", University of Texas Press.
Djorgovski S.B. 1986. in "Instrumentation and research programmes for small telescopes", eds Hearnshaw J.B. and Cottrell P.L., Kluwer Publ. Co., Dordrecht, p.243.
Djorgovski, S.B. & Dickinson, M. 1989. Highlights of Astron., 7, 645.
Evans, R. 1986. JAAVSO, 15, 25.
Grosbol, P. 1987. In "The optimization of the use of CCD detectors in astronomy", ESO conference n. 25, p.93.
Hearnshaw J.B. & Cottrell P.L. (eds) 1986. "Instrumentation and research programmes for small telescopes", Kluwer Publ. Co., Dordrecht.
Longo, G., Capaccioli, M. & Busarello, G. 1991. in preparation.
Longo, G. & de Vaucouleurs, A. 1983. University of Texas Monographs in Astronomy, 3.
Longo, G. & de Vaucouleurs, A. 1985. University of Texas Monographs in Astronomy, 3a.
Mackay, C.D. 1986. ARAA, 24, 255.
Meurs, E.J.A. 1987. in "The optimization of the use of CCD detectors in astronomy", ESO conference n. 25, p.105.
Ortolani, S. 1987. in "The optimization of the use of CCD detectors in astronomy", ESO conference n. 25, p.183.
Rufener, F. 1989. Highlights of Astron., 7, 519.
Spratt, H. 1986. Highlights of Astron., 7, 579.
Vigroux, L. 1987. in "The optimization of the use of CCD detectors in astronomy", ESO conference n. 25, p.237.
Walker, A. 1986. in "Instrumental and research programmes for small telescopes", eds Hearnshaw J.B. and Cottrell P.L., Kluwer Publ. Co., Dordrecht, p. 33.

Discussion

Walker: There is a problem here which has to be spelled out in detail. It applies both to the CCD work which you are advocating and to the JEAP which I discussed two days ago. These are complicated and expensive devices that rely heavily on software. There are few people or groups which can afford the $20,000 price of a CCD plus the cost of a 0.5-1.0 meter diameter telescopes. In my case I have got support from a charitable trust in England covering half the price of the JEAP. We will handle all the required software with the JEAP. If professionals want amateurs to take up work of this complexity, than they must be prepared to fund the hardware, provide all the software and some training in the use of the devices.

Longo: I am not fully aware of all possibilities offered by the amateur astronomers market. I know for sure, however, that many firms offer good quality CCDs (complete with all necessary software) at prices below $3,000. For what data reduction is concerned, there are already several packages (like for instance PC-

VISTA) available on personal computers and I expect many more to be released in the coming few years. I think that the real problem is that this type of program requires a dedication and an accuracy that very often are not within the reach of non-professionals. I agree that, if amateurs want to pursue these researchers, some training is needed.

Wisniewski: There is a tremendous need for long-term lightcurves of QSO's and active galaxies. If one can find a cosmetically clean part of a CCD and use only this part to do photometry of the object and of the comparison stars, flatfielding is not necessary for reaching photometric accuracy of a few per cent.

Cragg: Visual observations of supernova near galactic nuclei are much easier by visual methods than by others. Also, expense of CCD equipment usually precludes amateur participation. Supernova charts now enable magnitude estimates to be made at the time of discovery.

CHAPTER 4

ANALYSIS AND INTERPRETATION

VARIABLE STARS AND STELLAR EVOLUTION

John R. Percy
Erindale Campus
University of Toronto
Mississauga, Ontario
Canada L5L 1C6

Introduction

Evolution means **change**. Stellar evolution refers to the changes which occur in stars as they shine - or more generally as they lose radiant energy, mass and/or angular momentum. The time scale for evolution due to energy loss alone is proportional to luminosity/mass, averaged over the lifetime of the star. It is about 10^6 years for very massive stars, about 10^{10} years for the sun, and over 10^{12} years for the least massive stars.

Because these lifetimes are much longer than the age of modern astronomy (400 years or less), it follows that astronomers have not directly observed the normal, slow evolution of stars - only a few rapid phases like supernovae. There have been some attempts to compare modern photometric catalogues with the catalogue in Ptolemy's *Almagest* (Hertzog 1984), which is presumed to have been compiled by Hipparchus about 130 BC, but it is unlikely that significant evolutionary changes have occurred. The most interesting and controversial case is that of Sirius which, according to some interpretations, was red in antiquity. The weight of the evidence, however, is against this.

Variable stars are interesting and useful to astronomers because they provide information about stellar properties and processes - including evolution. The connections between variable stars and stellar evolution are so numerous and complex that it would be difficult to discuss them in a book, let alone a short article. You might want to browse through two recent conference proceedings: "Observational Tests of the Stellar Evolution Theory" (ed. A. Maeder & A. Renzini, D. Reidel 1984), and "Confrontation between Stellar Pulsation and Evolution" (ed. C. Cacciari & G. Clementini, ASP Conference Series #11, 1990).

Modelling Stellar Evolution

Theoretical "**models**" of stars are constructed by assuming that the star has a given initial mass and chemical composition, uniform throughout (very much like specifying a recipe for a cake). The physical behaviour of the atoms - the pressure, opacity, nuclear reaction rates - are assumed to be known as a function of temperature and density, from laboratory experiments and physical theory. The model must also obey four fundamental laws of physics, dealing with conservation of mass and energy, hydrostatic balance, and flow of energy. In principle, all this information is sufficient to predict the entire structure of the model, from its central temperature and density to its overall radius and luminosity. This information, in turn, can be used to determine how fast the star will lose radiant energy, whether it will cool and contract under its own gravity, or whether nuclear reactions will produce enough energy (and change the chemical composition of the star) to balance the energy budget of the star. The changes which will occur in the star in a given interval of time (a thousand years for instance) can be used to predict a **sequence of evolutionary models** which represent the changes in the star in its lifetime. These changes are most often displayed by graphing the luminosity and effective temperature of the model in the **Hertzsprung-Russell diagram**, and comparing them with the properties of real stars. Herein lies one of the most important uses of variable stars - to provide observational tests of theories of stellar structure and evolution.

Why Do Some Stars Pulsate?

Theory indicates that some stars will be unstable and will oscillate or pulsate if they lie in an **instability strip** in the H-R diagram. The gases in the outer layers of such stars act as a heat engine, absorbing radiant energy when compressed and releasing it when expanded, and thus converting the radiant energy into mechanical energy of pulsation.Stars spend most of their lives on the **main sequence** in the H-R diagram, because this is the locus of stars which are producing energy by the simplest and most efficient thermonuclear reaction - the transmutation of hydrogen into helium in their cores. The instability strip intersects the main sequence in a narrow strip between about A7 to F2 spectral types. Most stars of these types pulsate as **δ Scuti stars** - though often with vanishingly small amplitudes!

All other pulsating stars must evolve into the instability strip; that is, evolution must change their temperature and luminosity into values which result in pulsation driving and instability. It should then be possible to "explain" the existence and number of pulsating stars at each point in the instability strip; this would be one test of the theory of stellar evolution.

The test is not always straightforward. In the Magellanic Clouds, for instance, there are large numbers of Cepheids with short periods. This is due to the low abundance of elements heavier than helium in these galaxies. The evolution tracks differ from those in a galaxy such as our own with a more "normal" chemical composition, carrying more stars into the lower part of the instability strip where the periods are shorter.

The most impressive confrontation between stellar pulsation and evolution tracks occurs in the Magellanic Cloud cluster NGC1866, which contains large numbers of Cepheids, as well as many nonvariable stars. Comparison of the observed and theoretical H-R diagrams of this cluster provides a very stringent test of the detailed predictions of stellar evolution theory, and how stars become Cepheids.

The Period(s) of a Pulsating Star

A star, like any physical object, has natural periods of oscillation. The simplest is the **fundamental radial period**, which corresponds to an oscillation in which every part of the star expands and contracts in unison. This period P_f is given by

$$P_f = K_f R^{3/2} M^{-1/2}$$

where R is the radius, M is the mass, and K_f is a very weak function of R and M. Since stars in the same instability strip have about the same temperature, the luminosity L is proportional to R^2 by Stefan's Law, and there will be a relation between P_f and L:

$$P_f \propto K_f L^{3/4} M^{-1/2} T^{-3}$$

This is the famous **period-luminosity relation**, one of the fundamental tools for determining distance in the universe. If the period, luminosity and temperature of a pulsating star are known (which is often true if the star is a member of the galactic or globular cluster), then its mass can be estimated from the equation above.

Overtone Periods

Stars can pulsate in other radial modes. In the first overtone mode, for instance, there is a nodal sphere in the star which remains at rest, while the layers above and below pulsate in anti-phase. In the second overtone mode, there are two nodal spheres, and so on. The first overtone radial period is given by

$$P_1 = K_1 R^{3/2} M^{-1/2}$$

where $K_1 \sim 0.75 K_f$, and is also a weak function of R and M.

Some stars pulsate in two (or more) modes. **Double-mode pulsators** are common among δ Scuti, RR Lyrae and Cepheid variables. The two pulsation periods can be determined from long series of observations, giving P_f and $P_1/P_f = K_1/K_f$, both of which are functions of R and M. The values of these two functions can be determined by theoretical analysis of the natural periods of oscillation of model stars of various R and M, leading eventually to an estimate of these two quantities.

Astero- and helio-seismology

In some pulsating stars, many more than two modes are simultaneously present. On one hand, these stars potentially provide much more information about themselves. On the other hand, the information is much more difficult to decode. Some peculiar A stars, and some white dwarfs, pulsate with many periods in the range of a few minutes and, in principle, these periods could be determined from a single night's observations. Many δ Scuti stars are multi-mode pulsators, but their periods are difficult to disentangle, because they are of order hours in length. To obtain continuous series of observations, multi-longitude "campaigns" must be organized, as described by Breger in his article in these proceedings.

The most complex pulsator yet known is our sun. Observations of either the brightness or velocity of the solar photosphere reveal dozens of periods, corresponding to different modes of non-radial pulsation. In principle, each mode can provide information about different layers of the solar interior - their temperature and density structure and even their rotation.

Because of the complex mixture of pulsation periods, it is particularly important to obtain long and complete series of observations of our variable sun. Again, this can be accomplished through coordinated multi-longitude observations, and a special "global oscillation network" of small solar telescopes (the GONG project) is presently under development. In future, it may be possible to obtain similar series of observations of bright sun-like stars like α Centauri. **Helioseismology** and **asteroseismology** will provide as clear a picture of the interior of stars as X-rays and CAT scans provide of the human body!

The Cepheid mass problem

The story of "the Cepheid mass problem" is an excellent illustration of the interplay between observation and theory of pulsating stars. Several Cepheids are known to be double-mode pulsators, and comparisons between their periods and period ratios with those predicted by theory gave masses which were less than half the masses predicted by other methods (such as by comparing the stars' positions on the H-R diagram with the evolutionary tracks for stars of given masses). The solution to this "problem" was suggested a decade ago, but only verified recently: the values, used in the theory, of the opacity of the material in the outer layers of the stars were incorrect by a substantial amount. Only recently has it been possible to calculate these opacities more accurately; to do so has used thousands of hours of time on the most powerful supercomputers!

Period changes and stellar evolution

As a star evolves, its radius changes, and its pulsation period, being proportional to $R^{3/2}$, should change also. The change will be small because evolution is slow, but it should be observable in some pulsating stars because its effects are **cumulative**. Consider the analogy of three clocks : clock A which keeps perfect time every day, clock B which runs at a regular rate of 85999 seconds a day (instead of 86400), and clock C, which is running down by one second a day, so that its rate on successive days, is 86400, 86399, 86398, 86397... seconds. The errors of each clock, after 1, 2, 3, 4 and 5 days, will be : clock A: 0,0,0,0 and 0 seconds; clock B: 0,1,2,3 and 4 seconds; clock C: 0,1,3,6 and 10 seconds. The error of clock C, which is slowing down, accumulates as the square of the elapsed time, and will be very large after a few weeks.

To a first approximation, the radius of an evolving star will vary linearly with time, and the deviation of the star's behaviour from a constant period (called (O-C) or **observed** minus **calculated**) will be a parabola proportional to t^2. Many δ Scuti, β Cephei, RR Lyrae and Cepheid variables have (O-C) curves which are approximately parabolic. Actually, Fernie (1990) has recently shown that, in some cases, the evolutionary change in radius and period is expected to be quadratic, and the (O-C) curve is cubic; this is the case in the Cepheid Y Oph.

In other pulsating stars, however, the (O-C) curve is not parabolic. In some cases, it is a broken straight line, indicating an abrupt period change. In yet other stars, especially those with large radii and/or low masses, the (O-C) curves are irregular. In Mira stars, for instance, this is interpreted as being due to random fluctuations in period from one cycle to the next (see papers by Isles and by Lloyd in these proceedings). The same may be true in RV Tauri and related stars.

The Strange case of Polaris

Polaris, the North Star, is a Cepheid variable with a period of close to 4.0 days, and a small amplitude. It has been known for many years that the period is changing. The (O-C) curve is well fit by a parabola corresponding to a period increase of about 316 seconds per century. A more remarkable discovery is that the amplitude of Polaris is decreasing (Arellano Ferro 1983, Dinshaw et al. 1989), and may reach zero before the year 2000! An obvious explanation is that Polaris is evolving out of the instability strip, but Fernie (1990) has shown that the colours of Polaris (and of Y Oph, which may also have a decreasing amplitude) place them in the middle of the instability strip, not at the edge. The mystery remains .....

Pulsation and mass loss

Mass loss occurs in stars of almost every kind. In the sun and similar stars, "activity" leads to winds which remove a small fraction of the star's mass during its main sequence lifetime. In other parts of the H-R diagram, mass loss is more extreme. The most luminous stars, for instance, can lose mass at rates of 10^{-5} $M_\odot$/yr or more. Viotti; in his article in these proceedings, has described the mass loss in luminous blue variables such as η Car and P Cyg. The latter star varies consistently on a time scale of about 40 days, possibly due to pulsation. This pulsation may contribute to driving the mass loss.

Cooler hyperluminous stars undergo pulsation and mass loss. Rho Cas is the most prominent example. Its pulsation period is about 300 days, and its mass loss has both uniform and episodic components. Zsoldos and Percy (A&A, in press) have described the results of a long-term collaborative photometric study of this star, in which AAVSO photoelectric observers played an active role.

Pulsation and mass loss in Mira stars

The pulsation of Mira stars is described in many papers in these proceedings. The pulsation originates below the photosphere, in the regions where hydrogen and helium are partially ionized. This pulsation has a profound effect on the atmospheres of these stars, creating shock waves which travel outwards, "inflating" the atmosphere and increasing its density. In the cool outer reaches of the atmosphere, gases condense into dust grains. The outward-flowing radiation (mostly infra-red) from the star pushes on the dust grains, which carry gases along with them, initiating a stellar wind which may carry off 10^{-6} $M_\odot$/yr. Within a million years, these stars will lose enough mass to affect their evolution profoundly. In effect, they die <u>not</u> just by exhausting their thermonuclear energy, but by losing their mass.

Yellow Supergiant Variables

Several types of variables inhabit the upper central region of the H-R diagram, between the hot main sequence stars and the cool red giants and supergiants. The Cepheids have been mentioned; they are young, massive stars using helium as their energy source, evolving on time scales of millions of years. Their pulsation is relatively stable, aside from slow changes in period in some of the more massive stars, and the peculiar amplitude change in Polaris previously noted.

The other major group of yellow supergiant variables are older, less massive stars. They have previously gone through a **horizontal branch** phase of evolution, also using helium as an energy source. If their temperature placed them in the instability strip, then they pulsated as **RR Lyrae stars**. These stars are extremely useful for studying the properties and evolution of less massive stars, since they are found in large numbers in globular clusters, and have been studied over many decades especially by Helen Sawyer Hogg and her collaborators.

As horizontal branch stars evolve and expand, they pass through the instability strip as **BL Herculis stars**;, almost all of these stars show increasing periods, as evolution theory would predict.

As the star continues to expand into a red giant, it may occasionally undergo nuclear pulses or "burps" which bring it into the instability strip for a few thousand years as a population II Cepheid or **W Virginis star**. These evolve much more quickly than the young population I Cepheids; their period changes are large and irregular, and one (RU Cam) became almost non-variable in the 1960's.

At the end of the red giant phase, the star may undergo pulsation and mass loss as a Mira star, losing most of its "excess" mass and leaving a white dwarf core surrounded by a thin shell of normal material. The star shrinks quickly through the yellow supergiant phase until it becomes a hot white dwarf. As the star passes through the instability strip, it will pulsate in some fashion. It may pulsate rather regularly, in which case we would probably call it a W Virginis star. It may show alternating deep and shallow minima - the hallmarks of an **RV Tauri star**. Or it may be semi-regular or irregular, in which case it would be classified as an SRd star. Are these fundamentally different kinds of variables, or is there some underlying unity?

A fashionable new concept which may provide an answer is **chaos** - a form of irregularity which arises from very simple initial equations. Those who model stellar pulsation, using hydrodynamic techniques, have found that as one parameter (effective surface temperature) of the models is changed, keeping the others (mass and luminosity) constant, the behaviour of the pulsation changes from purely periodic, through alternating large and small cycles, to apparently irregular, thus reproducing the entire spectrum of observed behaviour (See Buchler 1990, for instance, for an overview and further references).

White Dwarfs

Virtually every star (except the most massive ones) ends its life as a core of degenerate matter, surrounded by a layer of normal gas. The core mass is roughly 0.5 $M_\odot$, and its composition is commonly carbon and oxygen. As the star ends its red giant phase, the outer layers of the star shrink, and the star passes quickly through the yellow supergiant phase. This should produce observable period decreases, but the actual period changes of RV Tauri stars, for instance, tend to be abrupt and erratic, though many decades of data on U Mon and R Sct do reveal period decreases which appear to be of the predicted order of magnitude.

The evolution of a white dwarf is simple in principle: it cools and fades at constant mass and radius. There are complications, however, because of the very thin layer of normal material around the degenerate core, and the possibility of gravitational/radiative diffusion in this thin layer.

Fortunately, some white dwarfs pulsate, though in complex mixtures of non-radial modes. Since the time scales are of the order of minutes, however, the periods can be determined very precisely. Already, they are providing "a wealth of otherwise unobtainable information about the structure and evolution of white dwarfs at various stages of their evolution" (Kawaler 1990).

Geometrical and Eruptive Variables

Most of this brief review is devoted to pulsating variables, partly because this is my bias, and partly because eruptive cataclysmic variables are well discussed elsewhere in these proceedings. The conference included an excellent review of eclipsing variables, but the manuscript of that paper was not available to us.

The evolution of binary stars is in principle simple, and in practice complex. If the two stars are much smaller than their separation, then they will evolve as two single stars. If and when their sizes become comparable with their separations, then they will gravitationally affect each other's evolution. They are also more likely to eclipse each other. Trimble (1983) has presented an interesting evolutionary classification of binary systems, based on the evolutionary stage of each component (pre-main sequence, main sequence, giant/supergiant, and collapsed object) and on their relative separation. For each class, one well-known example is given.

As in the case of pulsating stars, period changes in eclipsing binary stars can be determined, using the (O-C) method, from observed times of minimum brightness, and this is the main motivation of the AAVSO's eclipsing binary program, and similar programs of other variable star observing groups.

Rotation and Evolution

All stars rotate. Their rotation is derived from the random spin of the gas and dust cloud from which the star formed. Stars more massive than the sun rotate with equatorial rotational velocities of up to 500 km/s. At this velocity, the effective gravity at the equator is reduced by "centrifugal force", and matter can escape more easily from the equatorial regions of the star. This explains, in part, the mass loss in the **Be stars**, which gives rise to the emission lines from which their name is derived. The rotational velocities of the O, B and A stars remain constant throughout their relatively short lifetimes until the stars begin to expand into giants and supergiants. Conservation of angular momentum then causes the rotational velocity to decrease.

Roughly 10 per cent of the B, A and early F type stars have magnetic fields of up to several thousand Gauss. Chemical elements can diffuse upward or downward in the outer layers of these stars, under the combined influence of gravity (downward) or radiation pressure (upward), creating a star with a patchy "surface" and an unusual chemical composition. If these patches are non-symmetric relative to the axis of rotation, and if the axis of rotation is not pointed to the observer, then the star will appear to vary in brightness, colour and spectrum, with a period equal to its rotation period. Spectroscopically, these stars are called **Bp, Ap or Fp stars** and they are referred to as α^2 **CVn type variable stars**. These peculiarities seem to develop rather early in the stars' lives, and do not change much with age.

In stars like the sun, rotation is generally slower, but it plays a more complex and important role. Rotation produces a magnetic field by the **dynamo effect**. This magnetic field, in turn, produces sunspots and flares, and indirectly the chromosphere and hot corona. We see all these phenomena on other sun-like stars, especially on those which rotate more rapidly. The long-term monitoring of the brightness and chromospheric emission lines in dozens of sun-like stars has revealed some interesting generalities. There are periodic variations, on a time scale of days to weeks, due to the rotation of the stars. This enables the rotation period of the stars to be determined. In some stars, there are variations on a time scale of a decade, which are probably due to activity cycles like the 11-year sunspot cycle.

Observations of sun-like stars in clusters of known age (like the Hyades and Pleiades) reveal that young stars rotate rapidly, on the average. With time, their rotation decreases, as their weak stellar winds carry angular momentum outward along magnetic field lines. Specifically, the rotational velocity decreases as the inverse square root of the age of the star: $V_{rot} \propto age^{-1/2}$. The sun rotates very slowly, compared with younger stars of its mass. Its rotation (and activity) must surely have been much greater in its youth. The study of the variability of sun-like stars thus tells us much about the history of our sun's rotation and activity.

T Tauri stars.

T Tauri stars are sun-like stars in their extreme youth. They are found in large numbers in regions of star formation, such as around the Orion Nebula. Their photometric variations are complex and diverse. Some are very active, other less so. There are quasi-periodic variations on a rotational time scale, as well as longer-term variations which may be due to a variety of processes related to stellar activity. Visual, photographic or photoelectric monitoring of T Tauri stars is interesting and useful if it is done <u>systematically</u> (Herbst et al. 1987).

FU Orionis stars are a rare and extreme form of T Tauri stars; only a handful of stars of this kind is known. They are characterized by slow outbursts of up to several magnitudes, followed by an even slower decline. The cause of the outburst is not know, but an instability in an accretion disc around the star (akin to a dwarf nova eruption, but slower) is one possibility. Photographic monitoring of star-forming regions might help in discovering more of these stars. This, and long-term photometric monitoring of these stars would be a worthwhile activity for amateurs.

Supernovae

Supernovae are the best and most extreme example of the connection between stellar variablity and evolution, and of the wealth of information which a variable star can provide about its evolution. They are also a convincing demonstration that astronomy is an observational, rather than an experimental science. There is no shortage of articles and books on this topic (Arnett et al. 1989, for instance), thanks to the discovery and observation of SN 1987A.

This is the brightest supernova to be observed since 1604 AD, the first supernova whose progenitor was a known star, and the first supernova whose outburst of **neutrinos** was observed. Its present spectrum reveals the ashes of nucleosynthesis now visible, and its light curve and X-ray emission reveal the role of radioactive decay as a power source in the star. The metal deficiency of this supernova, and its consequent peculiar light curve and faint maximum luminosity make it a poor "standard candle" for determining distances to remote galaxies. Recently, however, HST and IUE observations of the circumstellar ring around the supernova have been used to derive a precise distance of 51.5 ± 3.1 kpc to the supernova and the LMC.

Evans, in these proceedings, discusses the visual search for supernovae. We must remember that SN 1987A was independently discovered by an amateur. Indeed the worldwide observation of SN 1987A with ground-based and space observatories is an excellent example of "International Coordination and Cooperation in Variable Star Observing".

References

Arellano Ferro, A., 1983, ApJ, 274, 755.
Arnett, W.D., Bahcall, J.N., Kirshner, R.P. & Woosley, S.E. 1989, ARAA, 27, 629.
Buchler, J.R. 1990, in Non-Linear Astrophysical Fluid Dynamics, ed. J.R. Buchler & S.T. Gottesman, Ann. New York Acad. Sci. Vol. 617, 17.
Dinshaw, N., Matthews, J.M. Walker, G.A.H., & Hill, G.M. 1989, AJ, 98, 2249.
Fernie, J.D. 1990, PASP, 102, 905.
Herbst, W., Booth, J.F. Koret, D.L. Zajtseva, G.V., Shakhovskaya, N.J. Urba, F.J., Covino, E., Terranegra, L.,. Vittone, A., Hoff, D., Kelsey, L., Lines, R., Barksdale, W. 1987, AJ, 94, 137.
Hertzog, K.P. 1984, MNRAS, 209, 533.
Kawaler, S.D. 1990, in Confrontation between Stellar Pulsation and Evolution. ed. C. Cacciara & G. Clementine, ASP Conference Series #11, 494.
Trimble, V. 1983, Nature, 303, 137.

Discussion

Boninsegna: During your report, you said that visual observations of Cepheids are of no value to establish an (O-C) diagram. Of course photoelectric data are better but don't you think that visual observations are better than no observation at all?

Percy: I suppose so. Observations of Population II Cepheids are particularly useful.

PERIOD ANALYSIS OF VARIABLE STARS

Jan Cuypers
Koninklijke Sterrenwacht van België
Ringlaan 3
B-1180 Brussel, Belgium

1. Introduction

In this paper a short description and some comments on the most frequently used methods of period analysis of variable stars are presented. Four main classes of methods can be distinguished: classic least squares, string length methods, phase dispersion minimization (PDM) and Fourier methods. I am aware that a lot of variants exist and I will make no attempt to describe them all.

The classic least squares method is well known by amateur and professional astronomers. The period adopted minimizes the sum of the squares of the differences between the observed and calculated moments of chosen phase. It is characterized by the O-C diagrams. General reviews are available as e.g. Willson (1986). String length methods require good sorting algorithms and are not so often used anymore. Description and references can be found in e.g. Cuypers (1987). Here I will only discuss the two other methods in detail. In both cases a test function is evaluated for the observational data at a series frequencies (or periods). A minimum or a maximum (depending on the method) indicates the frequency that is most probably present in the data. Remember that the test function will show asymmetric peaks if plotted against period.

If a frequency is considered present in the data and if the shape of the light curve is well known, it is recommended to make a (non-linear) least squares fit and determine the unknown parameters (amplitude, phase or other light curve characteristics) including the frequency itself as accurately as possible. This certainly has to be done when more than one frequency is necessary to describe the data. It can be achieved with software as described by Breger (1989) or others. See also Breger's paper on short-period variable stars in these proceedings. If the frequency search methods provide you with a good first estimate of the frequency, computation time for the least squares fit will be significantly reduced.

2. Phase Dispersion Minimization

The name Phase Dispersion Minimization or PDM was introduced by Stellingwerf in 1978 but the idea existed much earlier as e.g. in Jurkevich (1971). The phase (ϕ) is the decimal part of the number of cycles elapsed after a chosen epoch and allows to construct the well known phase diagrams. The phase interval is divided into *bins* and every observation is in at least one bin. In every bin the dispersion around the mean is small when you phase with the right period, but the dispersion could be very large if the period is completely wrong. You can also look at the mean values in each bin. If they vary a lot, the period is probably good. If they are all close to the overall mean, the period is not the good one.

To minimize the effect of an odd distribution of data points in the phase diagram, Stellingwerf (1978) introduced the *covers*: the phase interval is again divided into N_b bins, but this partition is carried out N_c times, each time $1/N_b N_c$ out of phase with respect to the previous one. The incomplete bins near phase 0 are supplemented with the corresponding parts near phase 1. The bins can be made large enough to avoid empty bins but a lot of bins are available to average and this results in a smoother test function.

It is not difficult to compute in which bin a data point falls: if you have N_b bins you can calculate a bin index: $J = \text{INT}(N_b \phi) + 1$ (1 is added to avoid numbering a bin zero) for every data point with phase ϕ. Then you can minimize the sum of the dispersions in the bins or maximize the squares of the bin means with the number of points in each bin as a weight.

3. Fourier Methods

In the second important group of period analysis methods are the techniques based on Fourier analysis. The classical periodogram is often used in astronomy and called the method of Deeming (1975). The test function at frequency f is

$$P_N(f) = \frac{1}{N}\left\{\left(\sum_{i=1}^{N} x_i \sin 2\pi f t_i\right)^2 + \left(\sum_{i=1}^{N} x_i \cos 2\pi f t_i\right)^2\right\}$$

with x_i observed at time t_i one of the N data points.

If the signal x is a sine wave with frequency f a high value of $P_N(f)$ will be seen at frequency f. The expression is very similar to the frequency dependent part of the residuals of a least squares fit of the data to a sinusoidal wave. If least squares fitting of observations with known mean (set equal to zero) resulted in a calculated set $x_i^c(f)$ the expression $\Delta R(f) = \Sigma_{i=1}^{N} x_i^2 - R(f)$ with $R(f) = \Sigma_{i=1}^{N}\{x_i - x_i^c(f)\}^2$, i.e. the sum of the squared residuals, is equal to

$$P'_N(f) = \frac{1}{2}\left\{\frac{\{\Sigma_{i=1}^{N} x_i \sin 2\pi f(t_i - \tau)\}^2}{\Sigma_{i=1}^{N} \sin^2 2\pi f(t_i - \tau)} + \frac{\{\Sigma_{i=1}^{N} x_i \cos 2\pi f(t_i - \tau)\}^2}{\Sigma_{i=1}^{N} \cos^2 2\pi f(t_i - \tau)}\right\}$$

This kind of periodogram was used by many astronomers among them Barning

(1963), Lomb (1976) and Scargle (1982). It is now generally known as the Lomb-Scargle periodogram. If r satisfies $\tan 4\pi fr = (\Sigma_{i=1}^{N} \sin 4\pi f t_i)/ (\Sigma_{i=1}^{N} \cos 4\pi f t_i)$, $P'_N(f)$ is independent from the chosen zero point for the time.

The test function $P'_N(f)$ has nice statistical properties. You can e.g. calculate the false alarm probability i.e. the chance that a peak of a certain height is caused by noise (Scargle 1982; Horne & Baliunas 1986; Cuypers 1987). It is indeed expected that when a large number of trial frequencies is examined, the chance of a noise peak due to an odd distribution of data points in the phase diagram increases. This has to be taken into account when identifying an intrinsic period in noisy data.

4. An Example and an Extension to Variable Periods

Horne and Baliunas (1986) gave a beautiful example of the power of period analysis. I constructed a similar data set but the times of observation (expressed in days) are the integers 1 to 100 plus or minus a randomly chosen value between 0.0 and 0.2. Ten points chosen at random were discarded. A sinusoidal wave with amplitude 1.5 (peak to peak) and noise randomly chosen between -1.5 and 1.5 was calculated. Although no periodicity is visible at first sight (see fig. 1), both the Lomb-Scargle periodogram (fig. 2) and the PDM statistic (fig. 3) show clearly the presence of the frequency 0.4 cycles per day (c/d). The false alarm probability is very low (less than 0.1%). The PDM method also marks subharmonics (one half, one third ... of the frequency) as significant, as known from theory (Stellingwerf 1978).

Note that due to the almost equal time spacing of the observations a symmetric pattern of peaks will appear in the region 0.5 to 1 c/d and the whole pattern will be repeated from 1 to 2 c/d, 2 to 3 c/d and so on. This characteristic known as *aliasing* is typical for astronomical time series and could make the identification of the real frequency very difficult and sometimes impossible. For more details about aliasing and possible remedies see e.g. Fullerton (1986) and references therein. A typical example in variable star studies is given in Cuypers (1983). Aliasing caused by the daily intertuptions of the observations is evidently avoided if a coordinated international (multi-longitude) campaign to monitor the star continuously is successful.

Many variants and extensions of the described methods exist. I will present only one extension that has been applied to AAVSO data. How are variable periods detected? Sometimes by comparing frequencies found in early and in recent observations but often the accuracy is not good enough for definite conclusions. Normally classical O-C diagrams are studied, but variability of periods can also be detected with Fourier or even better with PDM methods. Therefore define the phase in this way:

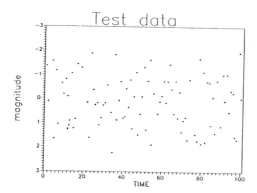

Fig.1. A series of simulated observations to illustrate the period analysis methods

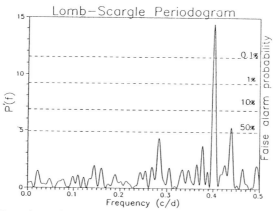

Fig 2. Lomb-Scargle periodogram of the data of Fig.1

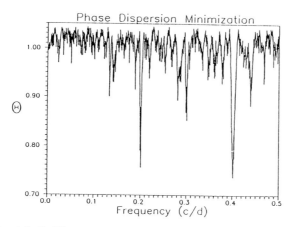

Fig. 3. Jurkevich-Stellingwerg test function (θ) for the data of Fig.1.

$$\phi(t) = [x(t)] = \left[\int_{t_0}^{t} f(t')dt'\right]$$

with x(t) the number of cycles elapsed since epoch t_0 and [x] the decimal part of x. For a constant frequency you will find the usual expression to calculate the phase, but for e.g. a linear time change of frequency you find $\phi(t) = [(t - t_0)f_0 + D(t - t_0)^2/2]$. Now you can extremalize the test function with respect to the frequency f_0 at epoch t_0 and the frequency change D.

To test the method I applied it to AAVSO data of T Cephei from 1927 to 1960. A constant period of 390 days was listed for this long period variable. However, a period decrease of 0.0024 days per day or 0.88 days per year was found with the extended PDM method. The resulting contour map (fig.4) and the accompanying phase diagrams (figs. 5 and 6) clearly show that a constant period was not the best solution for these data. The method has also been applied successfully to the ß Cephei star δ Ceti (Cuypers 1986), but remark that the method will only work properly when the period changes smoothly and the shape of the light curve remains the same during period change.

5. Comparison of Methods

Several detailed reviews, performance tests and comparisons of the different methods can be found in the literature. Fullerton (1986) reviewed these excellently. The most extensive tests on several of the described methods were performed and published by Heck, Manfroid and Mersch (1985). They and several others conclude that the methods are very similar in performance and efficiency, although not every one used all available information. I will only illustrate this and refer for the detailed conclusions to the papers cited above and the references in Cuypers (1987).

PDM methods were considered slow in computation. The first PDM programs however did not use the bin indices described above and this makes especially the Jurkevich method the fastest of them all. On the other hand, for the Fourier methods recurrence relations can be used (Ponman 1981; Kurtz 1985) and very recently a kind of fast Fourier has been developed to analyse very large datasets (Press & Rybicki 1989).

The well known properties of the Fourier methods are now much appreciated in the astronomical context. In criticising the PDM methods, one often forgets that the simple versions of the method (bins of equal length, no empty bins, no covers) also have well studied statistics (Jurkevich 1971; Heck, Manfroid & Mersch 1985, and recently rediscussed by Davies 1990). Only when using covers and special bin structures, statistics are not available.

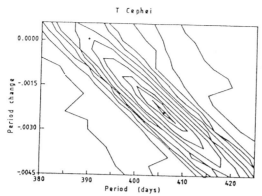

Fig. 4. Contour map with the most probable constant period (upper left) and the most probable linearly changing period (centre) found in the described AAVSO data of T Cephei.

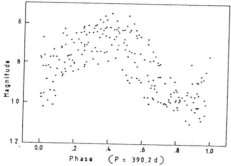

Fig. 5. Phase diagram with the most probable constant period in the described data of T Cephei.

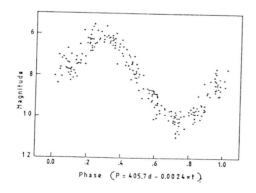

Fig. 6. Phase diagram with the most probable linearly changing period in the described data of T Cephei.

Swingler (1985) proved in a rather general case that PDM and Fourier methods bear more resemblance than originally thought. With this result in mind Swingler (1989) studied the local structure near the frequency peaks in the periodograms and concluded from simulations involving a single noisy sinusoid that PDM methods should be regarded as approximations to the Fourier method. He therefore recommends the Fourier methods as the technique of choice to analyze periods of variable stars. However, no extensive tests and studies are available for more complex situations such as data with non-sinusoidal signals, multi-periodic phenomena, variable periods, exceptional time distributed observations etc. Since at least in cases of light curves with unequal minima and/or maxima (as e.g. in Be stars) PDM methods seem to indicate more easily the presence of double or multiple waves, it is wise to use more than one method of period analysis.

Acknowledgement. I thank A.De Kersgieter for his help with the figures.

References

Barning, F.J.M. 1963, Bull, Astron. Inst. Netherlands, 17, 22.
Breger, M. 1989, Comm. Astroseismology 6,1, Vienna: Austrian Acad. Sciences.
Cuypers, J. 1983, A&A, 127, 186.
Cuypers, J. 1986, A&A, 167, 282.
Cuypers, J. 1987, Mededelingen van de Koninklijke Academie voor Wetenschappen, Letteren en Schone Kunsten van België, Academiae Analecta 49(3), 1-50.
Davies, S.R. 1990 MNRAS, 244, 93.
Deeming, T.J. 1975, Astrophys. Space Sci., 36, 137.
Fullerton, A.W. 1986, in The Study of Variable Stars Using Small Telescopes. ed. J.R. Percy, Cambridge University Press, 201.
Heck, A., Manfroid, J. & Mersch G. 1985, A&AS, 59, 63.
Horne, J.H. & Baliunas, S.L.1986, ApJ, 302, 757.
Jurkevich, I. 1971, Astrophys. Space Sci., 13, 154.
Kurtz. D.W. 1985, MNRAS, 213, 773.
Lomb, N.R. 1976, Astrophys. Space Sci., 39, 447.
Ponman, T. 1981, MNRAS, 196, 583.
Press, W.H. & Rybicki, G.B. 1989, ApJ, 338, 277.
Scargle, J.D. 1982, ApJ, 263. 835.
Stellingwerf, R.F. 1978, ApJ, 224, 953.
Swingler D.N. 1985, AJ, 90, 675.
Swingler, D.N. 1989, AJ, 97, 280.
Willson, L.A. 1986, in The Study of Variable Stars Using Small Telescopes, ed. J.R. Percy, Cambridge University Press, 219.

Discussion

Hall: To compute the "false alarm probability". is it necessary to know independently the uncertainty of an observed magnitude? Can you give me a reference?

Cuypers: No, it is even unjustified to use an independent estimate of the experimental error as shown by Horne and Baliunas (1986).

Isles: A comment on your analysis of period change in T Cep. I don't think that Fourier analysis or PDM are appropriate for testing period change in Mira stars. Using more appropriate tests on BAA data for this star in the years 1901-74, I find that this star is a marginal case. The methods used are described in Chris Lloyd's poster paper.

Cuypers: Fourier analysis is not appropriate, but the described PDM technique certainly works if a continuous period change is present. The data used span only a limited time range (33 years) and the smooth period mentioned here could be temporary behaviour of the star.

COMPUTER NETWORKS IN VARIABLE STAR ACTIVITIES

Veikko Mäkelä
Computing Centre
University of Helsinki
Teollisuuskatu 23
SF-00510
Helsinki, Finland

and

Aarre Kellomäki
Department of Biomedical Sciences
University of Tampere
P.O. Box 607
SF-33101
Tampere, Finland

The development of data processing was very rapid in the 1980's. Ordinary people have experienced this development in personal computers, which have gradually become useful devices e.g. for text processing. The relative prices of computers and accessory equipment have decreased, but technical properties (speed and memory capacity) have been improved. Thus, personal computers can use very sophisticated programs, and via telephone lines, they can communicate with each other or with larger computers.

At the same time, telecommunications have developed, enabling the establishment of fixed computer networks in industrial countries. Local networks have been connected to form larger international systems. The following computer networks, mainly academic, can be mentioned: INTERNET is originally an American combined system of some networks (ARPANET, MILNET, NFSNET) containing different domains. BITNET/EARN(European Academic and Research Network)/NORTHNET was originally a network between IBM computers, but nowadays other computers can use it also. SPAN (Space Physics Network) combines computers applied in space research. UUCP (Unix to Unix Copy) is a network for computers using a UNIX operating system. JANET is a mainly British network. FIDONET operates an amateur basis and uses general telephone lines. There are also commercial networks, e.g. packetnets run by national teleadministrations.

It is possible to communicate between different networks using gateway services. In most countries, the use of the academic networks is free for university teachers and researchers and, with certain restrictions, also for students. A private person must pay possible charges to the organization sustaining the network (typically $20 per month), plus the expenses of the telephone calls, which may be fairly high.

The three main applications of computer networks are:

1. Personal communication between users: The usual form is electronic mail, or "e-mail", a personal confidential electronic letter which is sent to the computer of the addressee, and can be read afterwards. On-line communications are possible, too.

2. NEWS-services are public correspondence. They contain many different activities:

   - "Mailing list" means sending the same e-mail to several addresses. It resembles subscription to a journal.

   - "News" is a more advanced system. Articles are distributed to the network, and they can be read anywhere in the world within a few hours. Because of the large number of people involved, the articles are divided in so-called "news groups", each dedicated to a particular topic. The news service is available in many networks, most notably in INTERNET and UUCP. Articles can be written by individual users without restrictions.

   - Conference systems" make it possible to arrange computer meetings or set up "bulletin boards" (BBS). Even connections to real conferences are possible.

3. FILE-services are means to transfer or utilize data files or programs. A "fileserver" receives orders of files and delivers the files to the computer of the orderer. Programs and data can be fetched directly from some computers, using a so-called FTP-program (file transfer protocol). "Data banks" are organized collections of data from which data can be retrieved using key words. Most data banks are commercial, so every data base query typically has an associated cost. BBS systems often include data which can be transferred using communication programs.

The following are some examples of possible and already existing ways of using computer networks.

Early warning circulars can be sent very quickly as e-mail or through news services. IAU sends this way plenty of information and notes about variable stars, e.g. about outbursts of novae and dwarf novae and fadings of R CrB stars. The most important notes are transmitted by national organizations to local networks, and the stars are often effectively observed already within a few hours after the warning. The early warning service of <u>The Astronomer</u> (England) should be mentioned in this respect.

Computer networks have been used for announcing international observing campaigns on interesting variables. Typically, a professional astronomer asks amateurs to monitor certain dwarf novae during a simultaneous scientific study and to inform the astronomer if an outburst occurs.

Observational results, both the original observations and the data files of associations, can conveniently and quickly be sent via computer networks. This has been done for years in our organization in Scandinavia. Nowadays we send results also to The Astronomer, and to AAVSO over the Atlantic Ocean. This application will become even more important in the future. Just now, we need international cooperation in order to plan an easy and reasonable system for saving the observations and transferring the data. All amateur observations could be collected by the AAVSO to form an enormous data bank which could be utilized by both professional and amateur astronomers.

For more information, please see the article: John S. Quarterman and Josiah C. Hoskins, "Notable Computer Networks" in Communications of the ACM, 29, 932, (1986).

Discussion

Dunlop: The extensive use of electronic mail has revolutionized the exchange of information in the United Kingdom in recent years. Alerts, detailed information and even charts are sent to most amateurs on a commercial network, which they use to report observations. To ensure fast communication with professionals, there are links to academic and scientific networks (SPAN, EARN/BITNET, etc.) through certain responsible amateurs. The instantaneous feed-back (in both directions) has proved invaluable.

Korth: I'd like to comment on the networks presented. For amateurs not working at a university or a computer firm it's difficult to have direct access to networks like BITNET or UUCP. Therefore Peter Blohen (Dahlenburg, F.R.G.) installed a bulletin board which now serves as a gateway and information system. On the national basis this is quite a convenient and reasonably-priced opportunity for amateurs to enter e-mail services. Maybe it will be possible in future to install a world-wide network of amateur host computers to allow a both cheap and fast exchange of information.

Shore: As an example, via SPAN, several important databases of satellite data can be accessed and studied by anyone in the world. Eventually HST and now with IUE, it is possible to study much of the spectroscopic data (after the proprietary period) to professional and amateur astronomers. The IUE data consists of about 60000 spectra of thousands of stars, galaxies and nebulae.

HIGHLY VARIABLE OBJECTS IN THE SOLAR SYSTEM

Wieslaw Z. Wisniewski
Lunar and Planetery Laboratory
University of Arizona
Tucson AZ 85721
USA

Introduction

We listened and talked at this AAVSO meeting about variability of stars and galaxies. Let us talk for a change about objects closer to us, objects in our Solar System. Asteroids and comets, those small solid bodies are no less interesting subjects for observation than variable stars. Often observations of them can be even more challenging, as the light reflected from those bodies is constantly affected by their perpetual motion and that of the Earth.

There is a pattern to the orbits of the planets described by what we call the Titus-Bode law (or rather not a law but a curious nature's coincidence). This "law" clearly predicted that at 2.8 AU from the Sun there should be a planet. Several German astronomers in the 18th century were searching, but the Italian astronomer Piazzi beat them to the discovery by finding on New Year's Day in 1801 the object which he named Ceres, for the Roman patron of Sicily. The object was at the expected distance from the Sun but relatively faint and therefore too small for a "decent" (expected) planet. German astronomers did not give up searching and soon discovered Pallas, Juno and Vesta. By 1890 there were 300 asteroids discovered; 1000 by 1923; 2000 by 1973. Today we are approaching 5000 numbered asteroids. Asteroids and comets are discovered as incidental objects during exposure of photographic plates or large CCD's, and by dedicated sky surveys.

Following the motion of a discovered object and obtaining several good astrometric positions can lead to orbit determination. Plotting positions of all known asteroids shows that the majority of them are in the rather large area between Mars and Jupiter, within what is known as a *main belt* between 2.2 and 3.3 AU from the Sun. This volume is so large that despite the existence of thousands of asteroids with diameters ranging from 1000 km down to 1 km, one could be in the middle of it and not see a single one. And yet on a geological timescale, most if not all asteroids have undergone collisions, breaking them apart. Since each object has its own defined orbit and this orbit is being constantly affected by perturbations, all those objects do not walk (move) together like good people on a sidewalk; some are moving in a more erratic way and have more chances of knocking against their fellows.

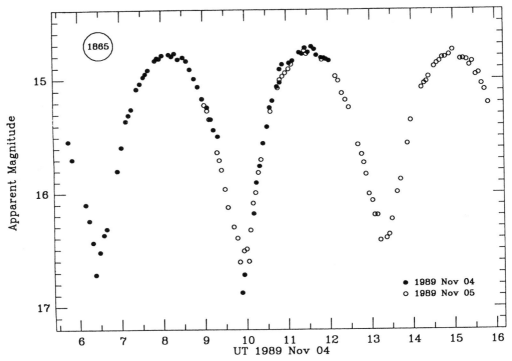

Fig. 1 - Light curves of 1865 Cerberus obtained on 1989 November 4 and 5. An example of a large amplitude of 2 magnitudes.

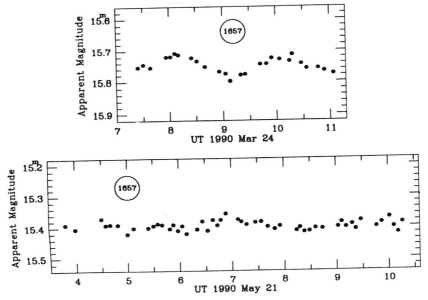

Fig. 2 - Light curves of 1657 Romera obtained on 1990 March and May 21. The small amplitude of 0.08 mag observed in March had disappeared by May 21.

There are several reasons why we study those small and seemingly negligible objects:

1) Asteroids and comets represent the only existing remnant planetesimals dating back to the formation of the Solar System. Perturbations by Jupiter perhaps led to the failure of planet formation in that region.

2) They are the most likely source for most meteorites, as collisional evolution pulverizes asteroids and delivers them to Earth.

3) There is a strong consensus today that collisions with Earth created conditions resulting in some species' extinction (e.g., dinosaurs).

4) We are approaching quickly a time when Earth-approaching asteroids will be a nonexpensive source of hydrogen and oxygen for space travel. The evidence is growing that some of the asteroids are extinct comets and therefore a reservoir of water. There are already well-organized institutes which study the engineering necessary for such exploration.

Observational Results

Naturally, the above-mentioned collisions change the asteroids' constitution and appearance, depending on actual conditions like asteroid size (mass), differential velocity and incidence angle of the collision. The effect of such a collision for a larger asteroid will be internal fragmentation (rubble pile model) when self-gravitation will prevent the escape of fragments (Farinella et al. 1982). Collisions breaking smaller asteroids apart will create a variety of shapes. As an irregularly shaped asteroid rotates about its axis it presents a varying amount of sunlight-reflecting surface area toward the Earth. Observations of the brightness of an asteroid over time will reveal a light curve showing two maxima per cycle. Matching features of the light curves obtained during the night or several nights will determine the period of rotation. This is the same procedure we apply when observing variable stars. While the geometrical relationship of the Sun, Earth and asteroid is constantly changing, we may observe different light curves for the same object. To disentangle the effects on light curves of shape and different aspects requires many observations at different epochs. The observations discussed below were obtained mostly with a photoelectric photometer on the 1.5 meter telescope on Mt. Lemmon near Tucson, Arizona. Those obtained with a CCD camera on the 2.1 meter Steward Observatory on Kitt Peak are indicated in the figure captions.

Some light curves have large amplitudes and some none at all. (Figs. 1 and 2). The lack of observed amplitude doesn't mean that there is none. As an asteroid moves over the sky its viewing aspect may change from equatorial to pole-on. Some asteroids are fast rotators of 2-3 hours (Fig. 3) and some are very slow over several days (Fig. 4). Please note that the time on Fig. 4 is in days, not hours. The

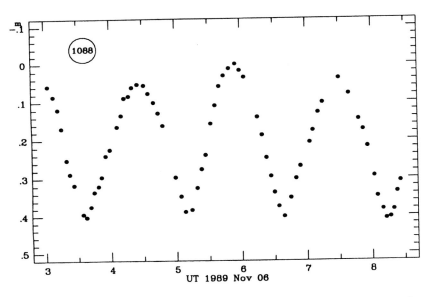

Fig. 3 - Light curve of fast rotator 1088 Mitaka obtained on 1989 November 6. The magnitude scale is adjusted to be 0 at maximum. A complete light curve was obtained in less than 3 hours.

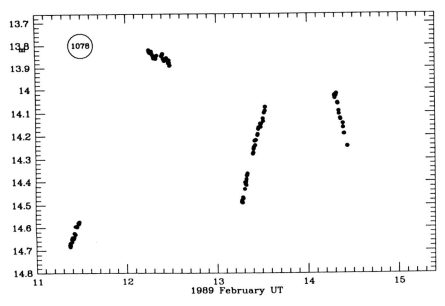

Fig. 4 - Light curve of slow rotator (P=3.6 days) 1078 Mentha obtained on 1989 February.

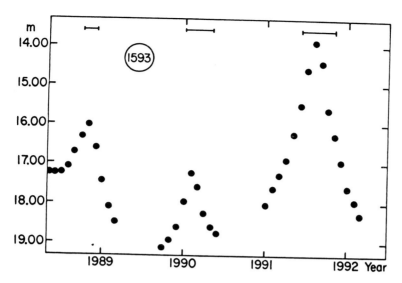

Fig. 5 Predicted brightness of Mars-Crosser 1593 Fagnes to demonstrate short observability windows.

analysis of lightcurves obtained over time leads to information on the asteroid's shape and its pole orientation in space.

Some not-yet-entirely-understood dynamical process is causing some orbits to become more elongated and eventually cross the orbits of Mars (Mars-Crossers) or Earth (Earth approaching asteroids) and even reaching as far as Venus and Mercury. Today the number of Earth-approaching asteroids is near 150 and every year more are being discovered. We call this group Apollo, Amor, Aten (AAA) depending on their semimajor axes and perihelion distances. AAA's occasionally get sufficiently close to collide with planets every million years or so. Erosion on Earth usually clears away all signs of such collisions in about half a million years. If not for this erosion, the Earth would look like the Moon. A very interesting book on the subject of collisions is by Chapman and Morrison (1989).

As clearly as AAA observations are of great importance, they are also the most difficult to observe. Fig. 5 shows the predicted brightness of Mars-Crosser 1593 Fagnes over several years. Bars indicate observability windows (not necessarily from the northern hemisphere). Our requirement in defining such windows is not only object brightness but a large solar elongation, allowing a number of hours of monitoring per night. As may be seen, those windows are often inconveniently narrow. Fig. 6 shows a history of light curves of 951 Gaspra. This asteroid will be imaged in October 1991 during a close flyby by the Galileo spacecraft as it travels to Jupiter. On Dec. 89 and Feb. 90 we note a flattening of one of the minima. Doesn't it remind us of a light curve of a W UMa star? Similar features are interpreted as such for some other asteroids (e.g. 44 Nysa). Lately this

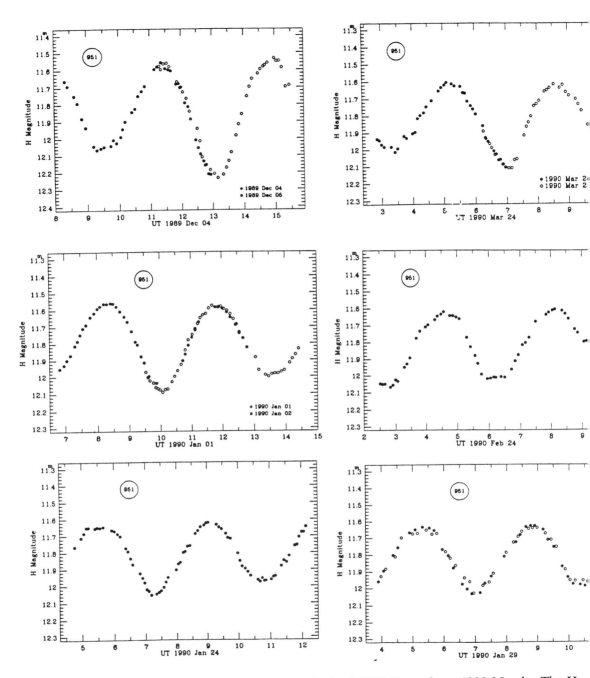

Fig. 6 - The lightcurves of 951 Gaspra obtained 1989 December - 1990 March. The H amplitude is the apparent magnitude corrected to unit distance. The January 1 and 2 observations were made with a CCD camera.

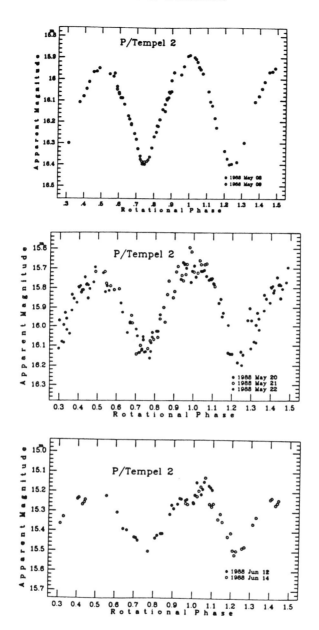

Fig. 7 - Light curves of comet P/Tempel 2 obtained on 1988 May - June. The derived synodic period of 8 hr 55.8 min and the same epoch, that of maximum on May 8.3646 was used to calculate rotational phase. The May 8 and 9 observations were made with a CCD camera. The visibly low noise on those dates is due to exceptionally good seeing and lack of cometary activity.

possibility got strong support from radar observations by Ostro et al. (1990) who convincingly have shown that the asteroid 1989 PB is a binary system.

Light curves similar to those for asteroids may be observed for comets. However, such observations are difficult and unpredictable as the comet must be nonactive for rotational variability to be detected. Otherwise the coma will be much brighter than the small nucleus and will bring the photometric amplitude to zero. After all, the projection of a typical 15 arcsecond diaphragm of a photometer at the distance of the comet will have a thousands kilometers diameter. Compare this with the few-kilometer diameter of the comet. In Fig. 7 a set of light curves of periodic comet P/Tempel 2 is shown. The comet was initially observed with virtually no coma. The decrease of the amplitude of the consecutive light curves is not caused by the aspect, as in the case of asteroids, but by slow development of the coma as the comet approaches the Sun. It is not possible to predict activity of the comet. That's why we know the period of rotation only for a handful of comets. It is my experience over the last 15 years to attempt monitoring of at least 10 comets per year and being able to observe rotation of a total of five of them.

Observational Techniques

Since this paper is addressed to experienced variable star observers, I shall discuss pertinent differences only. The observing technique would be virtually the same as for variable stars if not for the object's motion. This immediately implies that different comparison stars may be needed on different nights. Once I followed an asteroid from +70 degrees north to -60 degrees south. SAO stars, preferably of G type, are practical to use. If the objects do not move very fast, comparison stars may be selected somewhat ahead of its path, allowing the same star to be observed during several nights. Additional stars should be occasionally observed to check on comparison constancy. As we do not know if the weather will allow us to complete the observations, there is no need to bring comparison stars to the photometric system ahead of time. We may standardize them later for all asteroids we observed during an observing run.

The major problem will be, until we gain more experience, to find our objects on the sky. It certainly will be useful, however time consuming, to prepare finding charts with object positions marked every few hours. Unless our predicted position is in error, when identifying the field we should notice an extra star. It should move in several minutes. This is our asteroid or comet. With a well-pointing telescope we may save preparation time by finding the object by the motion itself. Presently there are several ways of obtaining asteroid positions. Each year the Institute for Theoretical Astronomy in Leningrad U.S.S.R. publishes positions and magnitudes for all numbered asteroids near their oppositions. Thanks to the availability of PC computers, it is possible to calculate ephemerides for any interval of time directly at the telescope. Frequently useful observing suggestions and ephemerides are published in the <u>Minor Planet Bulletin</u>, the quarterly publication of the Minor Planets Section of the Association of Lunar and Planetary Observers.

There is one major difference as compared with variable star observing - the sky measurements. As the object moves, so does the sky measurement location. We must be alert that faint stars are not in the photometer's diaphragm during the sky measurement. Then we should be aware that due to unavoidable scattering of light by the optics, nearby brighter stars may scatter enough light into the instrument's aperture to make a measurement unreliable. In other words, the asteroid will be measured on the background of uncontrolled scattered light. This is particularly true in the case of faint targets. It is not uncommon that we have to stop observing for a half hour or so to wait for the object to move. For this reason, unless we are really desperate, we avoid crowded regions (e.g. the Milky Way). For photoelectric photometry I would suggest selecting objects which are one magnitude brighter than we would normally do when observing variable stars. This will be compensated by the availability of CCD cameras extending the observable magnitude range.

A final note concerning instrumentation: as we are dealing with objects moving sometimes several degrees per day, it is advisable to have variable telescope tracking rates in declination as well. Otherwise slow moving objects should be selected. For more reading on observing techniques the chapter by R. Binzel in the book Solar System Photometry Handbook is strongly recommended.

Where do we publish observations of solar system objects? The main journals favored by professional astronomers are Icarus and Astronomical Journal. Amateur observations are published by the above-mentioned Minor Planet Bulletin.

Conclusions

Research on asteroids and comets is necessary to understand the beginning and evolution of our Solar System. This means that we must generate a statistically valid body of data on the physical characteristics of those small bodies, such as size, shape, period of rotation, albedo, color, etc. Despite many years of effort by many observers - professionals and amateurs - out of nearly 5000 numbered asteroids we have physical data, not necessarily complete, for less than 10% of them. So much is to be done by better-organized observations.

The range of magnitudes is such that it is not necessary to challenge the faintest one. There are many which can be successfully observed by telescopes smaller than 0.5 meter aperture. Because the light curve and brightness of asteroids and comets are so strongly affected by phase angle, distance, aspect, etc., observations to obtain the period of rotation should be made preferably on consecutive nights. Frequently, Earth-approaching asteroids or Mars-Crossers can be observed for a couple of weeks only, and are gone for ten, fifteen even twenty years before arriving again at favorable opposition (favorable opposition is when the object is not only at opposition but close as well).

object is not only at opposition but close as well).

It is virtually impossible to cover the whole light curve of a slow rotating asteroid from one observing site. One or two nights' observations from a few sites distributed around the globe will produce by far more complete data than hundreds of hours of observing at one site (weather permitting such a long run).

INTERNATIONAL COOPERATION IS VERY MUCH NEEDED AND APPRECIATED.

References

Binzel. R. P. 1983. "Photometry of Asteroids" in Solar System Photometry Handbook, ed. Russell M. Genet.
Chapman, C. & Morrison, D. 1989. Cosmic Catastrophics, Plenum Press. N. York.
Farinella. F., Paolicchi. P. & Zappala. V. 1982. Icarus 52, 409-433.
Ostro. S. J., Chandler, J. F., Hine, A. A.. Rosema, K.D., Shapiro, I. I. & Yeomans, D. K. 1990. Science 248, 523-1528.

CHAPTER 5

BLUE AND YELLOW VARIABLE STARS

THE STUDY OF SHORT-PERIOD VARIABLE STARS THROUGH INTERNATIONAL COOPERATION

Michel Breger
Institute of Astronomy, University of Vienna
Türkenschanzstrasse 17 A-1180 Wien, Austria

Introduction

The pulsating δ Scuti and β Canis Majoris stars have small light amplitudes near 0.01 mag, although a few δ Scuti variables can vary by as much as one magnitude. The periods range from about thirty minutes to eight hours. Usually more than one pulsation period is present so that the light curves do not seem to repeat. Both the small amplitudes and the presence of multiple periods provide a challenge to the observer. On the other hand, accurate photometric equipment on a small telescope together with careful observing and reduction techniques make observations of these stars very rewarding, since a considerable amount of scientific insight into the behavior of these stars as well as stellar evolution can be gained.

Who would have imagined that the careful investigation of the small period changes in selected evolved δ Scuti stars would detect period decreases, while the stellar evolution models predict period increases (Breger 1990)? Who could have predicted that in some cases more than six observable pulsation frequencies would be simultaneously excited in a star and that the amplitudes of pulsation would vary slowly and systematically from year to year? Another success concerns the observed period ratios of radially pulsating δ Scuti stars which, until recently, disagreed with the theoretically predicted ratios suggesting an increase in some critical stellar opacity values. The increase in stellar opacity values used for stellar models has now been comfirmed independently by calculations at Los Alamos Laboratory.

The International Observatory Network

Many short-period pulsating stars are excited by three or more periods simultaneously. Measurements for a week at a single observatory are usually insufficient to determine these periods because the daily observing gaps produce a large number of possible periods in the solution, among which the correct periods might not be recognized. One of the well-known problems caused by such regular observing gaps is one-cycle-per-day aliasing.

The aliasing problem leading to uncertain period determinations is illustrated in Fig. 1. A noiseless sinusoidal light variation with a frequency of 5 cycles per day

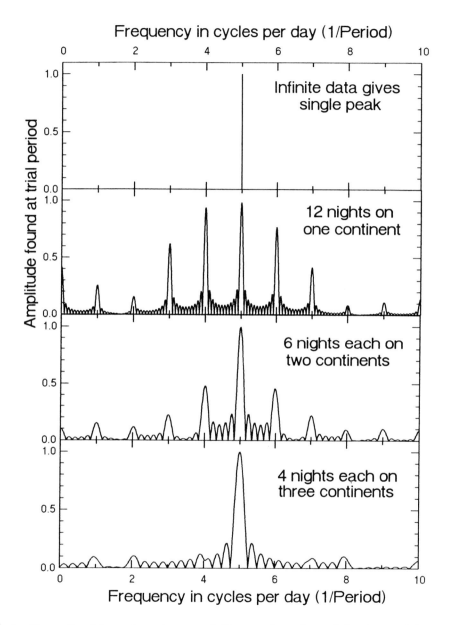

Fig.1: Example of how observing gaps influence the values of the detected frequencies. A noiseless signal with a frequency of 5 cycles per day is assumed (see top panel). Amplitude spectra are presented for 12 nights of observation, obtained at one or more observatories on different continents.

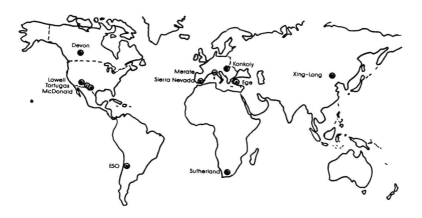

Fig 2: Location of the observatories cooperating with us since 1983 to study δ Scuti stars by interational photometric campaigns.

(period 4.8 hours) and unit amplitude was assumed. Not surprisingly, infinite data (top panel) give the correct answer. Amplitude-frequency diagrams are shown for twelve nights of hypothetical observations obtained at one, two and three observations in Europe, America and Asia. It can be seen that single-observatory measurements lead to problems with the period determination. The lack of symmetry in the structure and the unwanted amplitudes at very small frequencies are caused mainly by the fact that the average observed brightness does not correspond to the real average brightness of the star.

The value of international campaigns is shown in bottom two panels where the same number of observations lead to much improved period solutions.

The solution to this observational problem has been international cooperation with observers on several continents. In the Northern hemisphere a star is observed in Europe, the Americas and Asia for two weeks or longer. The variable weather conditions still lead to incomplete coverage, but this is not much of a problem since the most severe aliasing is caused by regular observing gaps, not the presence of irregularly spaced gaps. Even with measurements on only two continents the aliasing problem could still be considerably reduced.

The international campaigns by different groups have been guided by two different philosophies: (i) to persuade as many observatories as possible to observe a star, even if only a few measurements from a particular observatory can be obtained, or (ii) to carefully select a few observatories and astronomers to produce uniform data. The discussion of our experiences and possible pitfalls of campaigns discussed in the next section suggests that the second philosophy might be scientifically more productive.

Since 1983 we have coordinated eight international campaigns to analyze six δ Scuti stars. Astronomers at the following observatories have kindly participated in one or more campaigns: Xing-Long Observatory (China), Konkoly Observatory (Hungary), Ege Observatory (Turkey), Merate Observatory (Italy), Sierra Nevada Observatory (Spain), McDonald Observatory (USA), Tortugas Observatory (USA), Lowell Observatory (USA), Devon Observatory (Canada), South African Astronomical Observatory, and ESO (Chile). The location of these observatories is shown in Fig. 2.

These campaigns have enabled us to determine between four and seven simultaneously excited periods per star and, in some cases, identify them with specific nonradial pulsation modes. For some stars such as θ^2 Tau (Breger et al. 1989) and 4 CVn (Breger et al. 1990, 1991) the periods found could be confirmed independently from data obtained during different years, while for stars such as HR 2724 the multiple-period solutions probably are not complete.

The observational precision per single photometric measurement is usually around ± 0.002 mag, enabling the study of pulsation peak-to-peak amplitudes as small as 0.005 mag. In variable-star work this level of precision is not difficult to obtain with relatively inexpensive photometers and optimum observing techniques. As long as 25 years ago, we already obtained this relative precision with an old-fashioned direct-current photometer at Lick Observatory.

The key to the attainment of high precision comes from the fact that variable-star observations are relative measurements, since (usually) comparison stars are also observed. Consequently, many potential sources of error inherent in absolute color photometry cancel out. There exist a variety of publications discussing ways to obtain higher photometric accuracy, e.g. Young (1974), Borucki & Young (1984).

How to Organize Successful Multisite Campaigns

Our experience with campaigns leads to a number of recommendations that could assist in making campaigns successful. Our experience is, of course, influenced by the type of the stars that were observed by us, viz. small-amplitude variable stars with periods between one and eight hours. Before we list these recommendations in detail, an overall principle can be summarized as follows: Bad or uncertain data is worse than no data. The reason for this statement is quite simple: Since bad data cannot always be ignored (i.e. thrown out), enormous amounts of efforts would be expended to find (incorrect) solutions to fit the bad as well as the good data.

The different observations should be as homogeneous as possible. This requires complete planning before the international multisite campaign is started and the agreement by all the participating observers to adopt an uniform observing scheme. Needlessly to say, this requirement may present enormous difficulties of a diplomatic nature since most cooperations are based on good will only. The excellent article by Sterken (1988) on the coordination of astronomical observations contains discussions of other problems of diplomacy.

Below will be listed some recommendations for high-accuracy photometric measurements with single-channel photometers in detail.

Comparison Stars

The adverse effects of transparency and equipment variations are minimized by moving the telescope between the variable star and two comparison stars and interpolating the brightness during reductions. Two comparison stars are needed to check the constancy of the comparison stars and to obtain an estimate of the accuracy of the measurements. The importance of using at least two suitable comparison stars cannot be overemphasized. Furthermore, the comparison stars should be observed as often as the variable star. The aim is decrease the systematic errors as much as possible, not to produce beautiful-looking continuous light curves.

Considerable attention needs to be given to the choice of comparison stars. They need to be within a few degrees of the variable star, should not have completely different colors as the variable star, and be of similar brightness. However, since most stars are slightly variable, the two comparison stars cannot be chosen at random from their location in the sky. The candidates for comparison stars should, of course, not be known variables. Even if stars have not previously been examined for variability, the chances of undesired variability can be minimized by choosing comparison stars of specific spectral types or photometric colors. The presentation of variability statistics for all types of stars is beyond the scope of this paper. We have found stars of spectral types F2V to G0V to be excellent comparison stars for the observations of δ Scuti variables.

Observing Sequence

For highest accuracy the variations of atmospheric transparency as well as the astronomical instrumentation needs to be determined and corrected for in the data. Numerical experiments using data obtained by us at Kitt Peak National Observatory have shown that observing cycles as short as five minutes are required. The following sequence for single-star photometers is recommended: C1 - V - C2 - C1 - V - C2 etc., where V is the variable star to be studied and C1, C2 are the two comparison stars. For a 40-second to one-minute integration per star, each cycle takes about five minutes. Note that an alternate cyce of C1 - V - C2 - V - C1 - V etc., which provides a higher denisty of variable-star measurements, has been found to lead to lower accuracy and should only be used for stars with periods near half an hour.

If the periods of variation are shorter than 20 minutes (as in the case of the rapidly oscillating Ap Stars), it might be impossible to observe comparison stars at all. In this case the observer is required to use only extremely stable instrumentation and observe only at the best observing sites.

If the variable is observed nearly continuously with a comparison-star measurement made only every half hour, the results will probably be poor with the beautiful-looking light curves deceiving the observer into believing that high accuracy has been obtained. The observer has simply traded internal errors for high external errors hidden from sight. Such an approach is nevertheless useful if only the times of maximum light are wanted.

The previous discussion has assumed that errors from photon statistics can be neglected and more than 400,000 photons have been obtained for each measurement of 40 seconds, i.e. a count rate of at least 10,000 counts per second. This corresponds to a stellar brightness of approximately m_v = 9.0 for a 50 cm telescope with the Johnson V filter and 1P21 photomultiplier tube as detector. Should the stars be fainter, the observer might observe in the blue region (where the photomultiplier tube is more sensitive), use a more efficient photomultiplier, or lengthen the integration times.

Filters Used

From an astrophysical point of view, it is desirable to obtain measurements with more than one filter, while with a one-channel photometer single-color measurements are more accuate. Even worse, the desired second-order color information (such as amplitude ratios and phase shifts) needs to be of particularly high accuracy. The decision of whether of not to use more than one filter must be based on the expected sizes of the pulsation amplitudes as well as the astrophysical problem that should be solved by the observations.

It is extremely important that similar filters be used by all observers. The amplitudes of pulsation vary with wavelength and subsequent scaling of the measurements to convert from one color to another is an uncertain procedure to be avoided. On the other hand, the laborious and time-consuming procedure to observe transformation stars can usually be omitted, since transformations to a standard photometric system do not usually improve the variability parameters such as periods, amplitudes and phasing. We note here that it has been our experience that the broad-band Johnson V filter of the standard UBV system can be compared well with the narrow-band y filter of the Crawford/Strömgren *uvby* photometric system.

If only one filter is used, the question arises which filter is most useful. While measurements are usually made with the Johnson V filter, it could be argued that the B filter might be more useful, since the amplitudes of pulsation are larger by about 30%. This advantage is reduced by larger observational errors in the blue region. Especially for stars with small amplitudes near the noise level, the proper filter choice is crucial. One such example is the δ Scuti star HR 2724. For this star, unpublished four-color data for 19 nights by Balona are available and the variations in the brightness can be calculated before and after applying the multifrequency solution. If the detectability of pulsations is defined by the percentage by which the observed variations are reduced by the detected pulsations, a reduction of 100% would imply zero observational errors. For the four colors the following reductions of the variations in brightness are found:

ybv: 33%, *u:* 16%, *v:* 24%, *b:* 23%, *y:* 19%,

where *ybv* refers to the combined data of three filters and resembles single-filter observations. The 'best' wavelength region based on this data would be the blue region, between *v* and *b*. Furthermore, longer observations with a single filter would have led to the highest detectability of the pulsations.

Reduction Procedures

After the standard photometric reductions, for each observatory tables of times and magnitudes of the variable and comparison stars are available. To correct the variable-star data for the instrumental and atmospheric transparency variations, corrections are obtained from the comparison-star data by interpolation. The accuracy of the final data, however, does depend on how this interpolation is done. We adopt the following scheme:

(a) Light curves for the comparison stars are plotted. This immediately reveals those times where the agreement is poor or where the transparency is fluctuating rapidly. These times are marked for possible exclusion. Any variability of a comparison star will also be evident at this stage. These light curves of the constant comparison stars will also reveal whether the scatter in the brightness of the comparison stars is random or whether structure in the light curves is evident. This influences the choice of the interpolation scheme.

(b) The data of the two comparison stars are now combined by applying an overall zero-point difference for each observatory. We now have a brightness table for the combined comparison stars and the variable star.

(c) This step depends on the randomness of the scatter in the comparison-star data. If there exists very little random scatter and considerable structure, the difference variable-comparison is calculated by interpolating between four points (one of each comparison star at each side). However, if the scatter is essentially random in nature, this approach would increase the scatter in the variable-star data due to the additional measuring errors of the comparison stars. Consequently, a wider interpolation scheme involving eight or more points would then be more desirable. If a computer is used for these interpolations, one could experiment with different interpolation schemes. If a computer is not used, the accuracy might be even higher, if a smooth curve is drawn by hand through the comparison-star light curve with the structure of drawn curve reflecting the randomness of the scatter.

Relative Zero-Points of the Different Observers

After the conclusion of a multisite campaign, the measurements by the different observers need to be joined to each other. Since the same filter(s) and comparison stars are used, the brightness of the variable star relative to the comparison stars (the zero-point) should have a similar value for all the observing sites. Nevertheless, at the millimag accuracy level small differences in the zero-point between different observatories remain, which need to be determined and applied to the data. How can this be done?

The simplest method to determine the zero-point would be to average all the variable-comparison star measurements for each observatory and to ignore the fact that these averages are affected by the stellar variability. If a large number of observations are available, the stellar variability does cancel out.

If the large number of pulsation cycles can be covered in one night, it is possible to find the average brightness of these cycles and determine the individual observatory zero-points in this manner.

With the third method the astronomer applies approximate zero-point corrections, combines the data from the different observers, determines the pulsation periods and amplitudes and deduces the remaining systematic brightness deviations for each observatory from the overall solution. The analysis can then be repeated with the improved zero-points.

Two checks still need to be performed:

(a) Within each night, there should not exist zero-point drifts between the comparison stars. Such drifts at the millimag level could be the result of poor sky subtraction, incorrect coincidence corrections (for pulse-counting photometers), variable gain (for DC photometers), poor extinction coefficients or variable extinction, nonlinear tube response, or, of course, poor weather conditions. While one might state categorically that such drifts should not occur, they do happen. We experienced a drift during a few nights at McDonald Observatory due to photomultiplier tube fatigue! An artificial rectification of the data is, of course, a dangerous procedure.

(b) For each observatory the nightly zero-point differences for the two comparison stars can be determined. An agreement to $\pm$ 0.001 mag between the different nights should not be difficult to obtain.

If a particular observatory has only one or two nights of data available, the zero-point cannot be determined reliably. Furthermore, the two checks cannot be performed. The resulting uncertainty can have such a large adverse effect on the final results that these observations might need to be omitted. This is one of the reasons why multisite campaigns should be long enough in duration as to produce five or more nights of data per observing site. (The determination of reliable extinction coefficients is another reason.)

After careful reductions of the photoelectric measurements, the expected accuracy of the resulting brightness values of the variable star can be estimated from the scatter in the brightness variations of the comparison stars. It is nevertheless desirable to have an independent check of the light curves and the estimated values of the accuracy. This check can be obtained during an international campaign if some of the observations from different observatories overlap, which is often the case. For observatories spaced in longitude along the Earth, such overlap occurs when the program stars are setting at one observatory and are rising at another. In both cases, the airmasses at the times of overlap are relatively large and even the extinction coefficient may be changing (just after sunset). The comparison therefore occurs at those times when the measurements are least reliable and is, therefore, most valuable.

Fig. 3 presents simultaneous observations from two observatories obtained during a campaign of the variable star θ^2 Tau (Breger et al. 1989). The top of the figure shows the last measurements from the Sierra Nevada observatory in Spain (observer R. Garrido) and the first observations from McDonald Observatory (observer M. Frueh). The bottom part of the figure shows dual observations from Europe (observers R. Garrido and M. Paparo). A good agreement between these simultaneous observations is evident with a standard deviation of ± 0.002 mag per single measurement. This scatter is identical to the value of the accuracy of ± 0.002 mag determined separately for each observatory from the comparison stars. We can conclude that the reduced data can free of further 'hidden' errors.

Techniques of Period Finding

Modern computer techniques have replaced the traditional methods (usually involving phase and correlation diagrams) and make possible the determination of multiple periods. A small personal computer usually is sufficient and developed software is available (free) from researchers in this field of astronomy. Furthermore, pulsating stars with small amplitudes usually have sinusoidal light curves, which simplifies the techniques. Nevertheless the period-finding techniques may still be quite complex, but the complexities are hidden in the computer programs.

A bewildering variety of different methods of period-finding exists. Another talk at this meeting by J. Cuypers is providing a review so that only a summary

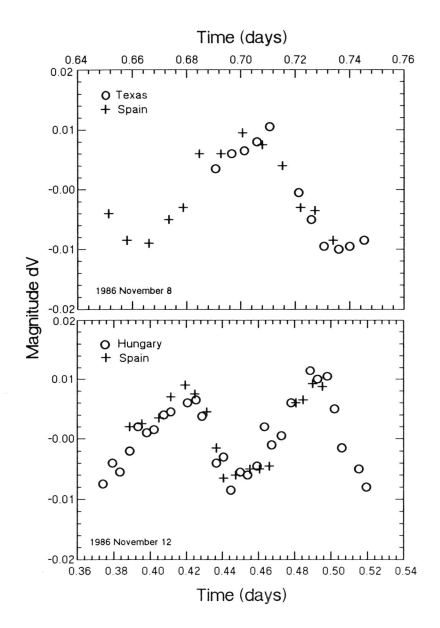

Fig 3: Comparison of simultaneous observations obtained during a campaign on the star θ^2 Tau.

containing our experiences is presented here. Three commonly used methods with variations are: (i) **FOURIER** (power or amplitude) spectra for unequally spaced data, which allow the determination of the frequencies with the largest amplitudes, (ii) phase dispersion minimization (**PDM**) which, as the name implies, shows the frequencies with the smallest scatter of the observed data in phase, and (iii) least squares of brightness residuals (**LSR**) which finds the frequencies with the smallest residuals between the final fit and the observed brightness (not to be confused with least squares methods that minimize the residuals in phase, see Cuypers 1987).

An excellent development of PDM was presented by Stellingwerf (1978). For the stars discussed here with sinusoidal or near-sinusoidal light curves, PDM provides identical answers as Fourier, but does not give the information on the expected aliasing caused by data gaps. Consequently, Fourier might be the slightly superior method, so that PDM will not be discussed here further. Many of the different methods used in the variable-star literature are variations of Fourier and LSR. Also, it is my impression that after years of developing their own methods, some groups of astronomers have actually been reinventing these excellent two techniques. As far as the CLEAN method is concerned, we are looking forward to a confirmation and a comparison of CLEAN results with those obtained by other methods for these stars.

A common feature of most period-finding techniques utilizing a computer is the calculation of a parameter describing 'the goodness of fit' for a large number of trial frequencies (= 1/period). For Fourier, the 'goodness of fit' is the size of the amplitude, for PDM the phase dispersion, and for LSR the brightness residuals. Most techniques require the user to specify a frequency range and the step size in frequency. A very common error made by a new user is the incorrect specification of the step size. A grid, which is too coarse, will not reveal the correct frequencies. The grid spacing is calculated from the time interval covered by the observations,

$$\Delta t = t(last) - t(first).$$

As was pointed out by Loumos and Deeming (1978), the maximum resolution of two very close frequencies is given by

$$\Delta f_{res} = 1.5 / \Delta t.$$

Of course, the frequency search grid should be much finer than given by the resolution. The ideal step size also depends on the possible presence of large gaps in the data. The best frequency should not be missed if the following step size in frequency is adopted:

$$\Delta f_{step} = 1 / (20 \Delta t).$$

Fourier Analysis

Two types of Fourier transforms are calculated. First, the actual observations in a time, magnitude array are evaluated within the desired frequency range with the appropriate frequency steps. If the data were of infinite length with no gaps, in the amplitude-frequency diagram the correct frequency would show a sharp amplitude peak (a mathematical δ function) at the correct frequency or frequencies. However, since data are not infinitely long and do contain observing gaps, the sharp peak is replaced by a complex alias pattern. The pattern is determined only by the times of observation (Deeming 1975). The computation of the shape of this pattern is the second calculation, called the *spectral window*, centered on a frequency of zero.

From a computation point of view, the Fourier method is an excellent and efficient method. Some computer time saving algorithms involving trigonometric recurrences exist. Some disadvantages relative to other methods need to be mentioned: (i) the use of Fourier requires that the zero-point of the magnitudes has been subtracted correctly, and (ii) the assumption that the frequencies with the largest amplitudes are the correct frequencies breaks down at those frequencies where the observing gaps (e.g. during day time) cover a significant part of the resulting light curve. For standard astronomical observations of short-period variable stars, the range between 0 and 1 cycles per day (periods longer than 1 day) may be overemphasized by Fourier methods. Furthermore, single-frequency methods such as Fourier and PDM perform very poorly if the data contains multiple periods, and have to use dangerous prewhitening (subtraction of frequencies) techniques to find the second, third etc. frequencies.

The LSR Method

This method using least-squares of brightness residuals does not suffer from these disadvantages. The technique determines the optimum frequency, amplitude, phase, zero-point and residuals at the same time by fitting the best sine curves to the data. In fact, for data with multiple periods a number of multiple periods can be fit simultaneously.

However, a major disadvantage of LSR is the very large amount of computer time required if LSR is used to scan all possible frequencies. Consequently, Fourier can be used for the overall scan and LSR is applied to only the frequency regions of interest. Special LSR computer programs exist which overcome the computer time problem by relying on approximate values of possible frequencies obtained by the efficient Fourier method. The exact value of the frequency (or multiple frequencies) is then also treated as an unknown to be mathematically calculated together with the other parameters of an optimum fit. For the last ten years, we have developed a period-finding computer package using LSR (program PERDET, Breger 1989). This program contains a very large number of additional features useful in analyzing astronomical data such as the zero-point checks described in the previous section.

Copies of the computer programs FOURIER and PERDET are available from the author for three computers (IBM PC, Atari Mega ST PC, and VAX) together with test data and handbooks describing the use of the programs. Some familiarity with the interpretation of Fourier power spectra and spectral windows (see Deeming 1975) is required.

References

Borucki, W.J. & Young, A.T. 1984, NASA Conf. Publ. 2350.
Breger, M. 1989, Comm. Astroseismology 6, 1, Vienna: Austrian Acad. Sciences.
Breger, M. 1990, Comm. Astroseismology 14, 1, Vienna: Austrian Acad. Sciences.
Breger, M. 1991, A&A., in press.
Breger, M., Garrido, R., Huang Lin, Jiang, S.-y., Guo, Z.-h., Frueh, M.& Paparo, M. 1989, A&A., 214, 209.
Breger, M., McNamara, B.J., Kerschbaum, F., Huang Lin, Jiang, S.-y., Guo, Z.-h. & Poretti, E.: 1990, A&A. 231, 56.
Cuypers, J. 1987, Academiae Analecta 49, 3, 1, Brussel: Academie voor Wetenschappen, Letteren en Schone Kunsten van Belgie.
Deeming, T.J. 1975, Astrophys. Space Sci. 36, 137.
Loumos, G.L.& Deeming, T.J. 1978, Astrophys. Space Sci. 56, 285.
Stellingwerf, R.F. 1978, A&J. 224, 953.
Sterken, C. 1988, in Coordination of astronomical projects in astronomy, eds. C. Jaschek and C. Sterken, Cambridge University Press, 3.
Young, A.T. 1974, in Methods of Experimental Physics, ed. N. Carleton, Academic Press, New York, 12A, 1.

Discussion

Sareyan: 1). I think several sky measurements have to be made according to the brightness of the variable and comparisons, to avoid systematic memory effects on the PMT, such as the "sky" after a bright star is ordinary higher than after a faint star.

Sareyan: 2). Another question: Could you please comment on the kind of compromise, we have to make between; distance from the variable and magnitude difference, when selecting the comparison stars?

Breger: 1). For acccurate work only photomultiplier tubes with small or zero memory effects should be chosen. If small memory effects do exist, the sky could be measured after every star observation. This would provide a uniformity of exposing the tube to light.

Breger: 2). Indeed, the choice of the best comparison stars must be a compromise and cometimes is quite painful.

Viotti: Monitoring of short period variables could alternatively be performed with observations beyond the Polar circles, such as at the Tromso Observatory in Norway (for more information please contact Mr. Van't Veer at the Institut d'Actrophysique, Paris) or in the Antarctic which is your opinion on this matter?

Breger: At certain times of year excellent, lengthy coverage can indeed be obtained at far Northern or Southern latitudes. However, the statistics from measurements obtained in the Antarctic may not have been too promising (e.g. due to the aurora australis).

PHOTOGRAPHIC PHOTOMETRY: APPLICATION TO POPULATION II VARIABLES

Martha L. Hazen
Harvard-Smithsonian Center for Astrophysics
60 Garden St.
Cambridge MA 02138, USA

In these days of efficient and linear electronic detectors, the question is often asked whether there is room in variable star work for the venerable technique of photography. In this paper, I will discuss astronomical photography and some of its characteristics and ramifications. I will also briefly describe Harvard College Observatory's unique and unrivaled historic collection of astronomical plates - a collection that I often refer to as "The Harvard Time Machine." Finally, I will take an admittedly rather chauvinistic view of some astronomical problems I am working on with photographic plates.

1. Photographic Photometry

Belserene (1986) has recently written a comprehensive and delightful view of astronomical photography, and the following remarks are intended not to parallel hers. The reader is urged to seek her paper out.

One can begin by asking: Is the technique of astronomical photography dead? For some applications, the answer is definitely *yes*; but for many variable star projects one can give an equally definite answer of *no*. Unlike visual, photoelectric, or CCD photometry, which are limited to a single star or a few very close stars at a time, photography can register a large number of stars over a wide field at a single shot. Until the advent of large format CCDs, then, photography keeps a unique place in astronomical photon recording.

However, there are a few factors/problems that one must consider when contemplating photographic photometry:

A. *Color Sensitivity:* Untreated ("original") photographic emulsions are sensitive over the ultraviolet, through violet and blue, to green. Modern emulsions can be made sensitive to other colors well into the infrared; however, they are still sensitive to the blue and ultraviolet. Therefore, it is normally necessary to use a filter to eliminate color sensitivity not wanted. Up until about 1950, astronomical photography generally employed unfiltered, blue emulsions; thus the amount of ultraviolet recorded depended on the optics and the atmosphere. Most variable star observations today are B-band: blue emulsion with a minus-UV filter. This is the simplest combination to use, and most variables show more variation in blue than yellow or red.

B. *Photometric Sequences:* Because photographic emulsions have a non-linear response to radiation (i.e., twice the light does not create twice the darkening of the plate), one must obtain a photometric sequence, or a selection of stars of known magnitude on the plate, preferably near the object being studied. Such sequences are not always easy to come by. Possibilities are:

1. Published sequences in the area - these are rarer than one might think!

2. Transfer from nearby published sequences - star clusters, or the HST Guide Star Photometric Catalog (Lasker et al. 1989), to suggest a couple of possibilities.

3. Do-it-yourself sequences - derived from your own photoelectric or CCD photometry.

4. Do without - one can pick a graded group of stars and assign arbitrary magnitudes; then, if the data warrant, magnitudes of the comparison stars can be determined after the fact.

C. *Interpolation devices and accuracy:* Some system is needed to compare the size/density of the image of the program star with those of the sequence stars. One very easy technique is simply to interpolate by eye, using a magnifier. With this technique, accuracy better than $\pm 10\%$ (0.1 mag) is hard to obtain; however, often this is all that one needs. It helps to have sequence stars at roughly half-magnitude intervals. The advantage of this technique is that it is extremely fast.

A second interpolation device is the flyspanker, perhaps now only of historic significance. A flyspanker is a series of graded star images on a photographic plate, preferably taken with the same telescope that was used for the plates being measured. One simply numbers the successive images, and estimates standards and program stars to 0.1 step along the flyspanker. A plot of magnitude versus flyspanker reading then becomes the calibration curve. This is one technique that requires practice, but an experienced observer can probably reach almost $\pm 5\%$ (0.05 mag) accuracy. Also, with this technique, transfers can be made over larger areas than with the eye-technique.

For the remaining two methods of interpolation, the accuracy limit is essentially set by the graininess of the plate. The inherent limits seem to be ± 0.15 mag for old plates, and ± 0.03 to 0.04 mag for modern emulsions.

The third interpolation device is the Variable Iris Photometer, where an adjustable iris is closed around an illuminated star image until the light coming through it is equal to that from some standard source. All the stars are measured, and a plot made of iris diameter versus magnitude of standard stars is applied to the program stars.

The final device is the microdensitometer. Here star images are scanned, and a calibration curve derived by plotting magnitude against the sum of the densities in the image. One needs a good software package to extract this sum. Note that this method is inherently no more accurate than the iris photometer; both are limited by the plate grain. The iris is faster for small jobs (perhaps fewer than 20 stars on a plate), and the microdensitometer faster for larger ones.

One final comment - the precisions quoted are for the best possible circumstances. Accuracy may be degraded by uneven development of the plate, variation of image quality across the plate, etc.

2. The Harvard Time Machine

If your variable is brighter than $B \approx 16\text{-}17$ mag, we may already have the data you need in Harvard College Observatory's collection of astronomical plates. In the Harvard plate stacks are some 400,000 plates, 1885 to the present. Perhaps 3/4 of these are direct (not spectra) blue plates, and the entire sky, north and south, is represented. The plates were taken with many telescopes (2.5 cm to 1.5 m apertures), at various plate scales (26 - 1200 arcsec/mm), with various field sizes (<1° - 40° on a side). For 18 mag > B > 15 mag, you will find anywhere from 2 or 3 to 100 images of a star for the past hundred years; if your star is brighter than 15 mag, there may be several hundred up to several thousand images.

The Harvard collection of plates is open to use by astronomers and also by qualified amateurs on request to the Curator of Astronomical Photographs, who currently is myself.

3. A Chauvinistic Look at Photometry and Population II Variables

Photographic photometry is well suited for studies of RR Lyrae stars (as well as other variables) in both galactic globular star clusters and the Magellanic Clouds, because there are usually many variables in one field. A summary of the characteristics of RR Lyrae stars can be found elsewhere (Hazen 1986). Briefly, RR Lyrae stars are pulsators with most periods between 0.3 and 0.8 day, absolute visual magnitudes of +0.6 (± an unknown amount!), and B-amplitudes of, usually, 0.6 to 1.5 mag. They are good distance indicators, and also provide clues to the late stages of stellar evolution.

Of the galactic globular clusters, some contain hundreds of RR Lyrae stars, while some contain few or none. The abundance of heavy elements in the cluster stars has something to do with this, but also there is apparently another factor, as yet unknown - the so-called "second parameter." Sawyer Hogg (1973) published a list of the known variable stars in globular clusters; this catalog revealed that many clusters had not been searched at all, or had been inadequately searched. So

I set out to work on some of the understudied southern clusters. To date I have published work on 20 clusters, some with large numbers of variables, and some with few or none, and I have plates for about 10 more. The plates were taken with the 1m Yale reflector at CTIO, and consist of series of 2 to 3 plates per cluster per night over runs of 7 to 14 nights. Variables are found with a blink microscope; measurements are carried out on an iris photometer. Some of the photometric sequences are from published material, but others I have had to derive from my own photoelectric observations. The results derived from these studies are distances to the globular clusters, and some data pertinent to the interrelationship among location in the galaxy, metallicity, and the absolute magnitude of these stars.

The work in the Large Magellanic Cloud requires a larger telescope since the mean B-magnitudes for the RR Lyraes is around 19.5 mag. The field I am working on is well out in the halo, around the globular cluster NGC 2210. Several colleagues have provided the plates and photometric sequence, chiefly my co-investigator Jim Nemec; most of the plates were taken with the CTIO 4m telescope. Blinking the whole 1° field took some 6 hours per plate pair; twenty pairs in all were blinked, with the discovery of more than 150 possible variables. The measurement was an ideal job for the PDS microdensitometer, since we had 100 standards to measure as well as the variables on each plate. There turn out to be some 20 to 30 RR Lyrae variables in the globular cluster (many of these had been discovered before), and about 50 new variables of this type in the halo field around the cluster. When the work is finished, we hope to be able to shed light on the depth of the LMC halo, the variation of absolute magnitude with metallicity, and the distance (yet another determination) to the LMC.

In summary, I wish to reiterate that photographic photometry is alive and well. And it is still true that, in some cases, photographic photometry techniques provide the best ways of proceeding - the study of RR Lyrae stars (population II variables) being one of the prime examples.

References

Belserene, E. P. 1986. In The Study of Variable Stars Using Small Telescopes, ed. J. R. Percy, Cambridge University Press., p. 43.
Hazen, M. L. 1986. JAAVSO 15, 201.
Lasker, B. M. et al. 1988. ApJS 68, 1.
Sawyer Hogg, H. 1973. Publ. David Dunlap Obs. 3, No. 6.

Discussion

Sareyan: *Over large fields, do you correct for differential atmospheric extinction? (In blue $K \sim 0.3$ to 0.5, so apparent magnitudes could be corrected to avoid systematic effects).*

Hazen: No, we don't. We try to make sure our sequences are reasonably near to the program star, so that the differential extinction errors are much smaller than the errors inherent in photographic photometry.

VISUAL DETECTION OF THE TWO PERIODS OF THE DOUBLE MODE CEPHEID EW SCUTI

A. Figer
GEOS
Promenade Venezia, 3
F-78000 Versailles, France

E. Poretti
Osservatorio Astronomica di Brera
Via E. Bianchi 46
I-22055 Merate, Italy

C. Sterken
Astrophysical Institute
University of Brussels (VUB)
Pleinlaan 2
B-1050 Brussels, Belgium

1. Introduction

Figer (1984), by analyzing visually-estimated magnitudes (from photographic plates taken between 1935 and 1939, Bakos (1950)), recognized that EW Sct is a double-mode Cepheid. The star had previously been classified as a classical Cepheid with a period of about 5 days (Bakos 1950). Eggen (1973), using photoelectric data, concluded that the period was about 10 or 11 days. Inspired by Figer's paper, Cuypers (1985) reanalyzed Bakos' and Eggen's data and determined the values of the two periods, i.e. 5.8195 days and 4.0646 days. The star was also included in the Long-Term Photometry of Variable Stars project (Sterken 1983) carried out at ESO, La Silla, and in the list of multiperiodic stars observed at Merate Observatory. From these measurements, Figer et al. (1991) determined *uvby* lightcurves, and confirmed the presence of the two periods.

2. Observations and Data Analysis

From August 1983 to November 1986 Figer (identification FGR) also obtained a homogenous series of 355 visual estimates of the star by means of Argelander's method and using a 106 mm reflector telescope located in downtown Paris. The long series of Figer's visual estimates was analyzed with the same least squares technique used by Figer et al. (1991) for the photoelectric measurements. The least squares period search procedure described by Antonello et al. (1986) was applied to the whole set of data. The method does not require any prewhitening as the amplitude and phase of each known constituent are recalculated for each trial frequency.

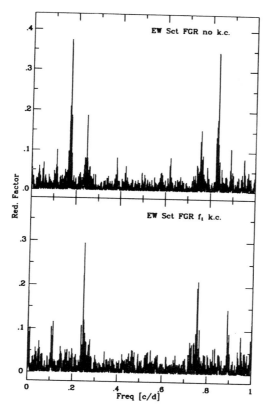

Fig. 1 - Power spectra of the visual estimates of EW Sct. Top: without known constituents (higher peak at $f_1 = 0.17164 \pm 0.00003$ c/d); bottom: with f_1 as known constituent (higher peak at $f_2 = 0.24591 \pm 0.00004$ c/d).

Fig. 1 shows the power spectra. The frequencies $f_1 = 0.17164$ c/d and $f_2 = 0.24591$ c/d are the two dominant peaks. The light curve associated with each period is very well determined, as can be seen from fig. 2. The rms residual of the fit is 0.14 mag, a reasonable value for visual estimates of a star located at a declination -7° by an observer located in Paris. The corresponding light curves to Bakos' data has less scatter (rms residual about 0.07 mag, see fig. 3), but their amplitude is only half the one observed photoelectrically or visually. The lower amplitude can be explained by the fact that, on the photographic plates, EW Sct was probably not resolved from DM -6°4816 a 0.3 mag fainter star at a distance of only 5 arcminutes.

3. Conclusions

Our results, based on a large number of first-quality visual estimates and on a proper analysis, illustrate very well the possibilities of an experienced visual observer, even if living in a "bright" site. They moreover show that interesting results can be obtained on members of classes of variable stars that are not commonly included in the observing programs of amateur astronomers associations.

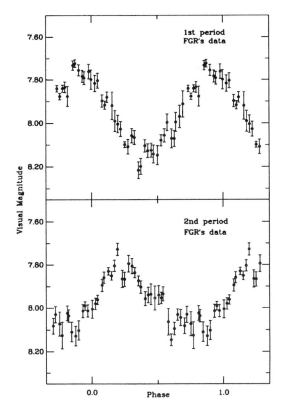

Fig. 2 - Light curves for the visual estimates. Top: light curve for the first period, bottom for the second period.

References

Antonello, E., Mantegazza, L., Poretti, E. 1986, A&A 159, 269
Bakos, G.A. 1950, Annales Sterrenwacht Leiden 20, 177
Cuypers, J. 1985, A&A 145, 283
Eggen, O.J. 1973, PASP 85, 42
Figer, A. 1984, GEOS NC 403, 1
Figer, A., Poretti, E., Sterken, C., Walker, E.N. 1991, MNRAS In Press.
Sterken, C. 1983, The Messenger 33, 10

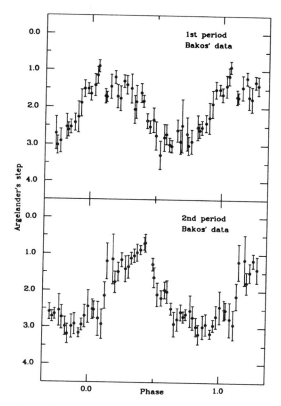

Fig. 3 - Light curves for Bakos' data. Top: light curve for the first period, bottom for the second period. The unit on the y-axis is the interval used by Argelander, and corresponds to 0.10 mag.

Discussion

Overbeek: How are the error bars determined in the diagram showing the light curve?

C. Sterken: The diagram is a phase diagram based on many cycles of variation. Every phase bin has several magnitude estimates associated with it, and this allows us to calculate the probable errors of each mean, which are the error bars given in the figures.

Samus: We have gathered about 20 good (±0.3 km/s) V_r values (by the CORAVEL technique) for EW Sct in May - July this year, and we hope to get much more. EW Sct is also being actively studied by Mr. G. Hacke (Sonneberg), mainly photographically, but he is also starting photoelectric observations in co-operation with Soviet observers.

ETA CARINAE, AG CARINAE, AND THE HUBBLE-SANDAGE VARIABLES

Roberto Viotti
Istituto Astrofisica Spaziale, CNR
Via Enrico Fermi, 21
I-00044 Frascati RM, Italy

1. The Hubble-Sandage Variables

Among the different types of variable stars there is a group of objects intrinsically bright enough to be easily identifiable in external galaxies. This group was first recognized in the pioneering work of Edwin Hubble and Allan Sandage who studied a small number of 16th to 17th magnitude variables in the M31 and M33 galaxies and described their large and irregular variations (Hubble & Sandage 1953, Fig. l). For this reason, this group of stars has been named "Hubble-Sandage Variables". In the following years, many other similar variables with m(pg) around 19 to 21 mag have been discovered in other galaxies such as NGC 2403 and Ho II of the M81 group, and in M101 (Tammann & Sandage 1968, Sandage & Tammann 1974a and 1974b, Sandage l983). Table 1 gives a list of HSVs. The overall behaviour of their light curves is the presence of irregular variations with an amplitude of 1 to 5 mag on time scales from months to decades. Variations with a smaller amplitude are also present on much shorter time scales. Generally these objects have a color index bluer at minimum, and present a marked ultraviolet excess. Spectroscopic observations are very difficult because of their extreme faintness. The still scanty observations have disclosed in many cases a rich emission line spectrum, sometimes resembling that of the galactic variable η Car (e.g. Tammann & Sandage 1968, Rosino & Bianchini 1973, Humphreys 1975, 1978, Gallagher et al. 1981).

The high intrinsic luminosity of the HSVs and the large amplitude of variability indicate that we are dealing with a powerful phenomenon quite similar to that observed in the brightest novae and in supernovae. But the nature of the phenomenon is far from being understood. In particular, it is unclear whether HSVs represent a homogeneous group of stars, or whether a different mechanism is working in each object. Unfortunately, the faintness of the objects presently makes their study difficult. Therefore, to understand the HSV-phenomenon it is of great interest to know whether similar objects are present in our neighbourhood. In the Large Magellanic Cloud (LMC) there is one star, S Dor, whose light curve closely resembles that of the HSVs. Its brightness of 9th mag at maximum clearly indicates a very high intrinsic (bolometric) luminosity of the star. Two other HSVs or "S Dor variables" have been found in the LMC: R127a and R71. In the Milky Way there are at least three stars, all in the Carina constellations, which can be identified as HS or S Dor variables: η Car, AG Car and HR Car. As we shall show below, P Cyg is another candidate HSV.

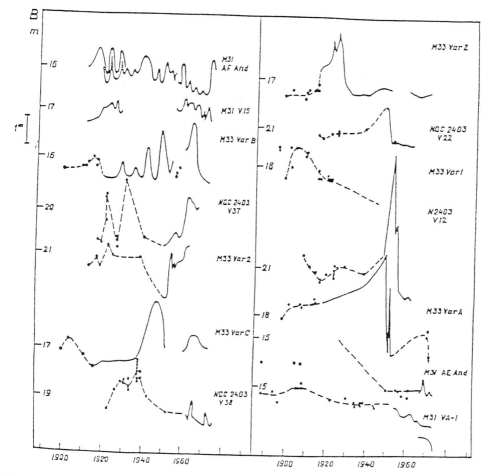

Fig. 1. The light curves of the Hubble-Sandage Variables (from Sharov 1976).

The luminous variables have been in recent times the subject of many investigations, and the main results have been presented especially during two international conferences held in Lunteren, The Netherlands in 1986 (Lamers & de Loore 1987), where it was agreed to call these stars "Luminous Blue Variables" (LBV) and in Val Morin, Canada in 1988 (Davidson et al. 1989). The reader can find most information in the proceedings of these conferences, as well ample references to past studies in the field. In the following, we shall first describe the behaviour of the HSV galactic counterparts AG Car, η Car and P Cyg. Then we shall compare them with the extragalactic HBVs and discuss possible interpretations of the phenomenon.

Table 1. Hyperluminous Variables in the Milky Way and Beyond (Veltroni 1984)

	star name	magnitude	finding chart
MW	η Car	6.2	
	AG Car	8.0	
	HR Car	8.1	
	P Cyg	4.8	
LMC	S Dor	9.5	HW67
	R 127a	10.4	HW67
	R 71	10.9	HW67
M31	AE And	16-18	Sh73
	AF And	15-18	HS53, Sh73
	Var A-1	16.6	RB73, Sh73
	Var 15	17.7	Sh73
M33	Var A	15.7	HS53
	Var B	16.2	HS53
	Var C	17.2	HS53
	Var 2	15-17	HS53
	Var 83	16.2	VB75, Sh81
NGC2403	V12	16.5	TS68
	V22	20.2	TS68
	V35	20.7	TS68
	V37	19.0	TS68
	V38	18.2	TS68
M101	V1	19.2	ST74
	V2	19.5	ST74
	V10	19.5	Sa83

References: HW67: Hodge and Wright 1967. HS53: Hubble and Sandage 1953. RB73: Rosino and Bianchini 1973. Sa83: Sandage 1983. ST74: Sandage and Tamman 1974. Sh73: Sharov 1973. TS68: Tamman and Sandage 1968. VB75: Van den Bergh et al. 1975.

2. AG Carinae

The peculiar spectrum and light curve of this star have long been known, but only in recent times has AG Car been the subject of detailed observations, including from space telescopes. One result after another has revealed the exceptional behaviour of this star, which clarified many aspects of the HSV behaviour, and the results represent a clue to understanding the whole phenomenon. We started to study AG Car some 20 years ago using archival spectra obtained at the Bosque Alegre Observatory, Argentina, and discovered large spectroscopic variations (Caputo & Viotti 1970). At our request, the former AAVSO Director M. W. Mayall

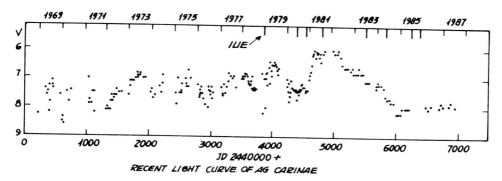

Fig. 2. The recent light curve of AG Car (from Hutsemekers and Kohoutek 1987).

(1969) collected the photometric data for the star, which were very useful to give a correct interpretation of our spectra. According to Mayall's data and to the recent ESO photometry published by Hutsemekers and Kohoutek (1987) (Fig. 2), until 1980 AG Car was irregularly variable between the 7th and the 8th magnitudes. In 1981 the star brightened to a historical maximum of V = 6. This was followed by a very unusual gradual fading from the 6th to the 8th magnitude which lasted about three years. Then AG Car has remained at minimum until the present. It is very difficult to understand this recent behaviour of the star on the basis of the photometric data alone. Fortunately, during this period AG Car was extensively studied especially from space in the ultraviolet. It turned out that, according to the optical and UV spectroscopic observations, the "temperature" of AG Car increased from around 10000 K to more than 30000 K. 1t should be however considered that in peculiar objects such as AG Car it is hard to define the temperature of the stellar surface. The above values are in fact referred to the mean ionization of the spectral lines and to the shape of the infrared to ultraviolet energy distribution. It was in fact found that during 1981-1985 the maximum of the radiation emitted by AG Car gradually shifted from the visual to the UV, so that most of its radiation is presently emitted in the invisible higher frequencies. Taking into account this fact, we have discovered that the total radiative power of the star has not decreased: the actual bolometric luminosity remained constant throughout the whole fading phase. This is not a new result, as it has been previously found in other astrophysical objects, including novae during their early post-maximum phases.

While the temperature was rising, the apparent radius of the star was decreasing, so that now AG Car appears much smaller than in 1981 (Fig. 3). Why this happened is one of the main topics discussed during the 1988 conference in Canada mentioned above. It is suggested that, like in the other hyperluminous stars, the atmospheric envelope of AG Car is unstable. A change of the gas opacity

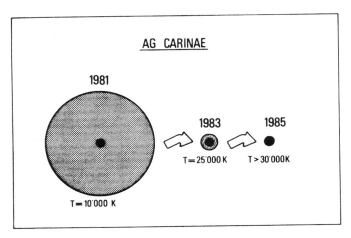

Fig. 3. Temperature and size variation of AG Car during 1981-1985. The large change was probably due to the external layers which became more transparent, or to their gradual shrinking.

and/or of the mass outflow may cause an expansion or contraction of the outer layers, followed by a decrease or increase of the stellar temperature. The observed time scales indicate that only a small fraction (say 1/10000) of the stellar mass was involved in the phenomenon. But the result was spectacular: the star apparently evolved from the white (A-type) supergiant category to the blue (B-type) supergiants, and finally to the Wolf Rayet stars in a few years. The same trip is covered by massive stars during advanced stages of their life, but in a much longer time. The point is that the displacement of AG Car in the temperature-luminosity diagram is not a stellar evolution. On the contrary it is a transient effect related to the instabilities of the external stellar layers. We expect the star to change again in the future, although we cannot anticipate when and how, and whether AG Car will undergo a more impressive event, such as a supernova outburst. Astronomy is beautiful for being unpredictable!

AG Car is most interesting also for a small circumstellar nebula. For the ring shape it closely resembles classical planetary nebulae (AG Car was in fact included in the Perek and Kohoutek Catalogue of PNs), with the not negligible exception that the central star is not a hot dwarf but a luminous supergiant. The nebula is expanding at a rate of about 50 km/s, suggesting that it was ejected some thousands of years ago (Viotti et al. 1988). Alternatively, the circumstellar matter, originating from a powerful mass loss during a previous evolutionary stage of the star, was in recent times swept away by the stellar radiation. This would imply that AG Car was previously much cooler than now, possibly a red supergiant. This hypothesis has been in recent times supported by the discovery of the presence of dust grains in the planetary nebula (Viotti et al. 1988; McGregor et al. 1988; Paresce & Nota 1989). The dust grains could have been formed in the cool wind of AG Car during its previous red supergiant phase. This hypothesis is however in conflict with some recent estimates of the absolute luminosity of the star (Humphreys et al. 1989). But this crucially depends on the distance of the star which still is uncertain. As we shall show below, dust might also be formed in the hot stellar wind. But this is not an easy way to solve the problem!

3. Eta Carinae

Eta Car is the most spectacular and fascinating HSV, and one of the most interesting objects in the sky. The star lies in a beautiful nebula, NGC 3372, which includes other peculiar objects, such as Wolf Rayet stars, and the hottest O-type stars identified in our Galaxy. But the region is interesting also in other spectral ranges. In fact, the X-ray picture of the field taken with the Einstein Observatory satellite is quite impressive for the many bright sources and the intense diffuse emission (Seward et al. 1979). In this regard, it should be noted that a gamma-ray source has been detected in the Carina region by the COS-B experiment. The infrared maps made with the IRAS satellite (Marenzi et al. 1990) show the very strong emission from η Car (actually the star is the strongest object in the sky in the mid-infrared, outside the Solar System), and another diffuse far-IR source centered on a dark lane about 10' NW of η Car (Harvey et al. 1979). The IR images also unveiled a large number of IR sources, indicating that the nebula is an active star-forming region.

Eta Car is close to a dark region resembling to some extent a river estuary. This part of the Carina Nebula had a particular role in the η Car history, which was recalled by Gratton (1963). One century ago, around 1830 to 1840, Sir John Herschel performed a detailed study of the southern sky. In particular, he made an accurate drawing of the Carina Nebula and determined the position of many stars in the field. According to his drawings, the central part of NGC 3372 was dominated by a close dark region, elongated in the NS direction with a central narrowing. The shape was similar to that of the (old fashioned) key holes, which gave the name of "Key Hole Nebula" to the Carina Nebula. Some decades later, other astronomers took a picture of this region of the sky but were unable to recognize the key hole as described by Herschel. In fact the central dark region appeared open in its southern part, as it appears in the present images. Gould (1871) was convinced that the change was imaginary and Herschel's drawings wrong. Gratton (1963) however noted that Herschel's measurements of the stars near the contours of the Key Hole were very accurate if compared with the present astrometric data, and it is quite hard to admit such a large error in his drawings. In addition one should consider that, at the time of Herschel's observations, η Car was very bright, a first magnitude star, while now it is many magnitudes fainter (at least in the visual). It is therefore easy to believe that when the star was bright, the nearby matter was illuminated by the stellar radiation. The fading of the star was like the shut-off of the lights illuminating the scenery in a theater. But the play has not ended!

The most puzzling aspect of the η Car phenomenon is the large light variation of the last century, which was described in detail by Gratton (1963). Being a southern star, η Car was not observed frequently enough to have a comprehensive light history. We know that, at the beginning of the 1800's, η Car was of magnitude 2-4 with occasional bright maxima in 1827, 1838 and 1843, when it attained the first magnitude. At the time of the 1843 outburst, η Car was the second brightest star in the sky after Sirius! Following a new short brightening in 1856, η Car underwent a spectacular light decrease from the first to the seventh magnitude in 14 years. Such a large fading was so similar to that observed in novae that it led some authors to include η Car in the category of novae (Payne-Gaposchkin 1957; Allen 1973). Indeed, this fading was not a real decrease of the radiative power from the star. In fact, as discussed above, η Car is very bright in the infrared. Its magnitude is 1.1 at 2.2 μ, -3.5 at 5 μ and -8 at 10 μ. The integrated magnitude from the infrared to the optical and ultraviolet is 0.0. This is the present "bolometric" magnitude of the star which is quite close to its luminosity during the light maximum of the last century (taking into account the estimated stellar color index at that time and the interstellar reddening; see Andriesse et al. 1978). At a distance of 2500 pc, η Car is 5 million times brighter than the Sun, one of the brightest stellar objects in the universe.

The conclusion is that η Car has probably not changed its luminosity. The radiation that the star was emitting in the visual in 1800, is now mostly emitted in the IR. This wavelength shift of the stellar radiation is due to a dense dusty envelope formed after the many outbursts of the mid-1800's. The dust has probably started to condense as a result of a change of the stellar wind structure. The envelope is presently absorbing most of the optical and UV radiation, and reemitting it at longer wavelengths. η Car is presently surrounded by a 10" irregular nebula formed by the matter ejected during the last ~ 150 years. The total mass of the nebula is around 10 solar masses, which implies a huge mass outflow from the star. But it also suggests a very high mass of the star itself (around 140 solar masses according to Andriesse et al. 1978). But many questions still remain open: what is the mechanism producing such a high mass loss? How is the dust condensing in the stellar wind? Is η Car really a very massive star?

Concerning the latter question, recent high-resolution observations have unveiled that η Car is a multiple system, with four objects, one significantly brighter than the others, separated by 0.1-0.2" (Weigelt & Ebersberger 1986). Thus η Car is like a small Orion Trapezium system. Yet most probably one component is the main (or unique) contributor to the huge mass ejection, and one - probably the same - is much brighter than the other system components. The multiplicity also suggests the hypothesis that the main component might be a close binary itself. In this case the luminosity and mass loss could have been powered by interactive phenomena in the binary system. A quasi-periodic modulation of the light curve with a period of ~ 3 years was noted by Feinstein and Marraco (1974), but not confirmed by other observers. This in another open question about our star. The nature of η Car still remains a mystery!

4. P Cygni

P Cyg is another hyperluminous star which must be included in the HSV group. P Cyg belongs to the Cyg OB 1 association at a distance of 1800 pc. Its effective temperature is 19000 K and the absolute luminosity, taking into account the interstellar reddening, is 0.7 million times the solar one (Lamers et al. 1983), which is in the range of the HSV luminosities. This star, well-known for its peculiar spectrum, characterized by emission lines flanked on their violet side by absorption lines, is presently nearly stable around V = 4.9. But during the 17th century, the star underwent large luminosity variations between the 3rd and >6th magnitude (Lamers 1985). Thus its past behaviour was in many respects similar to that of the HSVs and can be interpreted following the above discussion of AG Car. During the light maxima, P Cyg was most probably cooler than now (in fact the star was noticed to be red), while when its visual magnitude dropped, the stellar temperature was much higher. It is also possible that, like in η Car, these fadings were caused by formation (followed by dissipation) of dusty envelopes. There is, however, another consideration. P Cyg remained at constant magnitude for more than two centuries (Lamers 1985). Long periods of "quiescence" have also been observed in other HSVs (e.g. Var C in M33, Fig. 1). Therefore observations even covering several decades might not be sufficient to unveil most of the HSVs in a given galaxy: many of them might be in a long lasting "quiescent phase" with some small amplitude variability similar to that observed in other luminous blue stars. Thus at the moment we cannot decide about the frequency of HSVs in galaxies: the present number of identified objects could be largely underestimated.

5. What Are Hubble-Sandage Variables?

We have shown that the galactic HSV AG Car displayed large photometric variations, but the stellar luminosity remained constant. A similar behaviour was found in the Large Magellanic Cloud variables S Dor, (which evolved from WR to B-type), R127a and R71 (Wolf 1989). In the case of η Car a large luminosity fading was caused by dust grains formed in the stellar wind after the paroxystic phase of the mid-1800's. Actually, circumstellar dust has been found in AG Car and in some blue supergiants in the Magellanic Clouds. Thus dust is an important ingredient in the HSV phenomenon. We believe that most of the large light variation of the Hubble-Sandage Variables is due, not to nova-like "outbursts" and violent ejection of matter, but to change of the external structure of their atmospheric envelopes.

Occasionally, the conditions in the outer envelopes are such as to favour the condensation of dust. This may cause large luminosity fadings like those observed in Var A in M33 (Fig. 1).

The Hubble-Sandage (or Luminous Blue) Variables certainly are massive stars whose outer layers are highly unstable. Their luminosities are around one million times that of the Sun, and their masses in the range from about 40 to more than 100 solar masses. But the mass estimate should be revised if they are binary or multiple stars. Also their frequency is unknown for they might be in a quiescent phase for a long time. Once again the new observations have raised new questions instead of solving the problem!

Finally, what will be the fate of these stars? In some cases it has been announced that objects like η Car and AG Car will "soon" explode as type-II supernovae. Being massive objects, this is their expected fate. But to anticipate the time of their outburst is mere science fiction. Fortunately, Nature has always succeeded in disavowing the expectations of the Astrophysicists! The most famous example is represented by the recent Supernova 1987A in the Large Magellanic Cloud. For a long time, theoreticians had been expecting the next supernova explosion in our Galaxy after Kepler's supernova of 1604. Finally in 1987 the big event took place! But instead of in our Galaxy, the supernova exploded in the nearby LMC. Well, never mind! The Large Magellanic Cloud is well known for the abundance of peculiar objects, and SN 1987A is close to the very rich 30 Doradus (or Tarantula) Nebula. This region includes an enormous number of peculiar hypergiants, and Wolf-Rayet stars, which are especially concentrated near the core of the Nebula. But observations indicate that the star that actually exploded, (the Supernova Progenitor), was not so peculiar. In fact it practically escaped astronomers' attention until the explosion! And now we are in trouble to understand why a stupid common blue supergiant exploded without giving us a signal, a warning of its coming fate. So, do not expect that puzzling variables such as the HSVs will easily unveil their nature. We only have to continuously observe them.... and wait their next move.

References

Allen C.W. 1973, "Astrophysical Quantities", 3rd Edition, The Athlone Press, London.
Andriesse C.D., Donn B.D., & Viotti R. 1978, MNRAS 185, 771.
Caputo F. & Viotti R. 1970, A&A 7, 266.
Davidson K., Moffat A.F.J., & Lamers H.J.G.L.M. (editors), 1989, "Physics of Luminous Blue Variables", Kluwer Acad. Publ., Dordrecht.
Feinstein A. & Marraco H.G. 1974, A&A 30, 271.
Gallagher J.S., Kenyon S.J., & Hege E.K. 1981, ApJ 249, 83.
Gratton L. 1983, in "Stellar Evolution", L. Gratton editor, Academic Press, New York, p.297.

Gould B.A. 1871, MNRAS 32, 16.
Hodge P.W. & Wright F.W. 1967, The Large Magellanic Cloud, Smith. Pub. 4699, Smith. Press, Washington.
Hubble E.P. & Sandage A. 1953, ApJ 118, 353.
Humphreys R.M. 1975, ApJ 200, 426.
Humphreys R.M. 1978, ApJ 219, 445.
Humphreys R.M., Lamers H.J.G.L.M., Hoekzema N., & Cassatella A. 1989, A&A, 218, L17.
Hutsemekers D. & Kohoutek L. 1987, A&AS, 73. 217.
Lamers H.J.G.L.M. 1985, "Luminous Stars and Associations in Galaxies" IAU Symposium no.116, C. de Loore et al. eds., Reidel, Dordrecht, p.157.
Lamers H.J.G.L.M., de Groot M., & Cassatella A. 1983, A&A 128, 299.
Lamers H.J.G.L.M. & C.W.H. de Loore (editors) 1987 "Instabilities in Luminous Early Type Stars", Reidel, Dordrecht.
Marenzi A.R., Ranieri M., Viotti R., & Baratta G.B. 1990, "The Infrared Spectral Region of Stars", C. Jaschek & Y. Andrillat eds., Cambridge University Press, Cambridge.
Mayall M.W. 1969, JRASC, 63, 221.
McGregor P.J., Finlayson K., & Hyland A.R. 1988, ApJ, 329, 874.
Paresce F. and Nota A. 1989, ApJ, 341, L83.
Payne-Gaposchkin C. 1957, "Galactic Novae", North-Holland.
Rosino L. & Bianchini A. 1973, A&A, 22, 453.
Sandage A. 1983, AJ, 88, 1569.
Sandage A. & Tammann G.A. 1974a, ApJ, 191, 603.
Sandage A. & Tammann G.A. 1974b, ApJ, 194, 226.
Seward F.D. et al. 1979, ApJ, 234, L55.
Sharov A.S. 1976, in "Variable Stars and Stellar Evolution", Reidel, Dordrecht, The Netherlands, p.275.
Sharov A.S. 1981, Peremenye Zvezdy 21, 485.
Tammann G.A. & Sandage A. 1968, ApJ, 151, 825.
Van den Bergh S., Herbst E., & Kowal C.T. 1975, ApJS, 29, 303.
Veltroni C., 1984, "Le Variabili di Hubble-Sandage", unpublished thesis, University of Padova.
Viotti R. et al. 1984, in "Future of Ultraviolet Astronomy Based on Six Years of IUE Research", NASA CP-2349, p.23.
Viotti R., Cassatella A., Ponz D., & Thé P.S. 1988, A&A, 190, 333.
Wolf B. 1989, "Physics of Luminous Blue Variables", K. Davidson et al. eds. Kluwer Acad. Pub., Dordrecht, The Netherlands, p.91.

Discussion

Shore: There are a number of both galactic and extragalactic luminous blue variables, and their relatives, which should be continually monitored. Their importance for the study of late stages of massive star evolution is enormous. For northern hemisphere observers, the star HD316285 = He 3-1482 is especially important. Unfortunately, most *of these stars are known only in the southern hemisphere. The LMC stars R126 and R127, and S Dor, are especially important. And of course, I echo the need for monitoring HR Car and η Car. Such activity is enormously important for us poor satellite types, who want to study the changes in mass loss in the ultraviolet. This is especially critical given the operation constraints with HST, and the need to schedule observations some weeks in advance.*

RECENT WORK ON R CORONAE BOREALIS STARS

David Kilkenny
South African Astronomical Observtory
P.O. Box 9
Observatory 7935, South Africa

1. Introduction

The 87th I.A.U. Colloquium held in Mysore, India in late 1985 (Hunger, Schönberner & Rao 1986) was the first international conference dedicated to hydrogen-deficient stars and included a number of substantial review papers covering both the theoretical and observational aspects of research into the small but fascinating group of stars known as the R Coronae Borealis (RCB) variables. The purpose of this paper is to describe briefly the characteristics of RCB stars and to review the more recent observational work on them and the related cool hydrogen-deficient carbon (HdC) stars.

RCB stars exhibit the following properties:

1) Unpredictable deep minima, apparently due to obscuration by ejected material, in which the star fades by up to 8 or 9 magnitudes in V with often a relatively fast fall in brightness and a slower recovery to maximum light.

2) Very hydrogen-deficient and rather carbon-rich atmospheres. Where detailed analyses exist, other elements seem to have near solar abundances (see Lambert 1986).

3) They appear to be either F-G supergiants or carbon stars with luminosities probably in the range $-3 < M_v < -5$ (Feast 1975).

4) Where detailed photometry exists, RCB stars show small-amplitude variations; in the case of RY Sgr, at least, these are apparently due to radial pulsation (Alexander et al 1972). In all RCB stars the variations are not strictly periodic, might perhaps be multiperiodic, and can appear quite irregular. In some cases, the radial velocity variations are very small and may show the same periodicity as the light variations (e.g. Raveendran et al 1986).

5) All RCB stars have infra-red "excesses" due to circumstellar dust (Glass 1978; Kilkenny & Whittet 1984; Walker 1986). These excesses can vary with amplitudes ~1-2 mag at L and on time scales of ~1000 days or more (Feast 1986, Menzies 1986) and can be explained as resulting from circumstellar dust shells with typical temperatures ~500-900°K (Walker 1986). The small amplitude stellar variations are apparent at near infra-red wavelengths and neither they nor the infra-red excesses are affected by the occurence of deep minima. Thus neither the star nor the circumstellar dust shell are seriously affected by deep minima (Feast 1986).

6) With one possible exception, the RCB stars appear to be single stars; V482 Cyg might be a member of a quadruple system (Gaustad et al 1988) which would imply $M_v \sim -2.8$.

7) On the basis of distribution and radial velocities, the RCB stars appear to belong to an old disk or possibly galactic bulge population (Drilling 1986).

The hydrogen-deficient carbon stars are spectroscopically very similar to the RCB variables but do not exhibit deep minima or have significant infra-red excesses. Until recently they were often referred to as "non-variable RCB stars" but it has been shown that they do have small amplitude variations (Kilkenny et al 1990).

A list of known RCB and HdC stars is given in Table 1, which is based on Drilling & Hill (1986). XX Cam has been observed in deep minimum only once in almost a hundred years (Yuin 1948, Rao et al 1980) and that was only ~1.7 mag deep. The star exhibits small amplitude variations but has no infra-red excess and so probably should be considered to be an HdC star (Rao et al 1980). It could easily be that the other HdC stars undergo rare deep minima which have simply not been observed so far. LR Sco and V605 Aql are not RCB stars (see Table 2); NSV 6708 clearly is a "new" RCB star (McNaught & Dawes 1986, Kilkenny & Marang 1988, Lawson & Cottrell 1989 and see McNaught 1988 for the story of the repeated discovery of this star!) UX Ant, V517 Oph and FH Sct are Hydrogen-deficient and probably are RCB stars but photometric evidence for deep minima is mostly lacking.

Table 2 is a list of some RCB suspects which have been shown not to be RCB type variables. It is hoped that these stars can be removed from the RCB literature.

2. The small amplitude variations (pulsation?)

As noted previously, RY Sgr is believed to be a radial pulsator but for the other stars the situation is unclear; some do not show obvious radial velocity variations (though data are often sparse) and non-radial pulsations or multiple frequencies might be present. It might also be that some small obscurational minima are mixed in with the small amplitude pulsational variations near maximum light. Holm & Doherty (1988) have suggested this possibility in an attempt to model IUE/visible region observations of R CrB itself during a small (~0.5 mag) "dip" in 1982. The theory of pulsation in hydrogen-deficient stars has been reviewed by Saio (1986) for example.

In the "best behaved" RCB star, RY Sgr, Kilkenny (1982) and Marraco & Milesi (1982) showed that the pulsation period was decreasing in a way which is consistent with rapid evolution - the star appears to be contracting on a time scale of a few tens of thousands of years - at very close to the rate predicted by Schönberner's models for a star of ~0.7 solar masses with a C/O core and He shell (Schönberner 1977, 1986; Weiss 1987). Recently, however, Lawson & Cottrell (1988, 1990) have suggested that a series of piecewise solutions fit the data better than a simple quadratic ephemeris and until the mechanism behind this period change is understood, some doubt must remain in the interpretation of the period decrease. Indeed, as Feast (1990) has pointed out, if changes in the decrease rate can occur on time-scales of 10-20 years then the overall period decrease, measured over ~100 years, cannot be interpreted as an evolutionary effect with certainty.

For the other RCB stars the situation is even more confused. For R CrB itself, Goncharova et al (1983) found multi-periodicity at ~27, 40 and 54 days, whilst Fernie (1989) found evidence for periods of ~27, 44 and 74 days from 1971 data but only one period, 43.8 days, from 1985-7 data. Most recently, Fernie (1990) found that the persistent 44^d period had disappeared in 1988 to be replaced by 59, 31 and maybe 23 day periods! It is clear that the small amplitude variations (pulsations??) in R CrB can fade from ~0.3 mag in amplitude to much less than 0.1 mag (e.g. Fernie 1982) and that a simple, single period is not the whole story. Add to this the possibility that small obscurational minima might be confusing the period analysis and some idea of the difficulty of observing RCB stars can be obtained. Lawson et al (1990) used Fourier analysis to derive a number of periodicities, sometimes multiple, for most of the southern RCB stars but the baselines of the data are not enormously long compared to the period (especially if there are multiple frequencies) and the amplitudes are often small so that some uncertainty must remain in the interpretation. There are reasonable grounds for accepting a double period for RY Sgr (37.7 days, ~0.18 mag; 55.2 days, ~0.09 mag), which is an additional reason for worrying about the evolutionary interpretation of the period decrease, and the derived periods for the other stars range from that of RY Sgr to ~227^d (Y Mus) with substantial concentration around 40 days. (See also Jones et al 1989, Marang et al 1990).

The hydrogen-deficient carbon stars were first shown to exhibit small amplitude variations by Kilkenny et al (1988) who suspected the periods to be ~20-40 days. Lawson et al (1990) find periods in the range ~40-160 days with no real evidence for multiperiodicity except in HD 175893 where periods/amplitudes of ~41 days, 0,07 mag and 20.6 days, 0.07 mag were found; the latter looks very much like an harmonic of the former, however.

The only thing which is really clear about the small amplitude variations is that more data are needed; baselines of several years with rather good observation density will obviously be necessary when periods are > 40 days. Radial velocity monitoring would also be a decided asset.

3. Dust formation, deep minima and circumstellar shells

The theory of formation of the dust which is believed to cause the deep or obscurational minima in RCB stars has been reviewed by Pugach (1984), Fadeyev (1986), Feast (1986), Goldsmith & Evans (1988), Efimov (1988) and others. Plausible models for the minima are based on the Loreta-O'Keefe model of ejection of carbon-rich material; Feast (1986) showed that "puffs" of material ejected in a random direction within a cone of semiangle ~20° could maintain the infra-red excess from the circumstellar cloud and produce the observed frequency of deep minima with a mass loss rate of ~10^{-6} solar masses/year. This would require puff ejection about every 40 days or so - close to the observed pulsation period in, for example, RY Sgr and other RCBs. It is tempting, therefore, to link the ejection of puffs from the star with the pulsation period, indeed Pugach (1977) and Goncharova et al (1983) have noted that for RY Sgr and R CrB there appears to be a link between "pulsation" phase and the onset of deep minima. Percy et al (1987) however, have analysed a larger set of data for R CrB and find no relation between period ~30 - 50 days and the onset of deep minima, even allowing for slightly decreasing periods. In addition, there is no plausible mechanism for ejecting material in a blob or puff with a radial pulsation, (a shell or disk should be ejected) and the models by Fadeyev (1988) indicate that dust formation takes place more than 20 photospheric radii from the stellar surface (at ~7000°K) although he suggests temperature changes in the pulsation cycle might cause dust formation in an outflowing carbon-rich medium.

Recently, a number of detailed observations, mostly of R CrB itself, have reinforced the Loreta-O'Keefe model and have shed some light on the dust formation; it is particularly interesting to see how the satellite data (IRAS, IUE, ANS) can enhance our knowledge of the processes involved. IUE data for R CrB and RY Sgr have been used by Hecht et al (1984) and Holm et al (1987) to infer from extinction curves that the ejected puff of material is composed of glassy or amorphous carbon particles with sizes ~10-50 nm (100-500Å). The former authors suggest that the apparent absence of graphite could be due to grain formation in an hydrogen-deficient medium. In this context, Wright (1989) has postulated fractal grains, clumps of spherical carbon grains of size ~50Å, as an explanation of the observed ultraviolet extinction features.

IRAS data have revealed the existence of a huge cool shell, apparent at 60 and 100μm, around R CrB and a similar but smaller shell around SU Tau (Gillett et al 1986; Walker 1986; Rao & Nandy 1986). The shell around R CrB is ~8pc in diameter and therefore around a hundred times bigger than the largest known OH/IR source or planetary nebula. Although very cool, ~25-30°K, and uniform in temperature, the interstellar radiation field cannot provide the energy to heat the shell. Feast (1990) has suggested that impulsive heating from R CrB could heat the shell without causing a significant temperature gradient in it. Gillett et al (1986) tentatively identify the shell as the "fossil" remnant of the hydrogen-rich envelope of the star ejected smoothly over a period from ~150000 to ~25000 years ago (though there are no data on the actual chemical composition of the shell).

Back on the ground, Rao et al (1990) have reported enhancement of C_2 and CN in the 1986 minimum of R CrB; this molecular enhancement might be an intermediate stage in the formation of grains. Cottrell et al (1990) obtained the most detailed spectroscopic and photometric data to date on the onset of an RCB deep minimum during the 1988 decline of R CrB. These data and earlier results from declines of RY Sgr, NSV 6708 and UW Cen lead the authors to describe "red" and "blue" declines, defined by the colour (UBV) behaviour in the early stages of decline. Their interpretation is geometric; a "blue" decline occurs when the ejected "puff" expands to cover the photosphere and chromosphere progressively, a "red" decline occurs if dust obscures the visible photosphere and lower chromosphere on formation.

Extensive JHKL infra-red photometry of ten of the southern RCB stars is currently being analysed at the SAAO. The SAAO infra-red group have monitored these objects for about 15 years on a ~monthly basis to derive information about the long-term behaviour of the infra-red excess. Comparison of 3J(1.2μm) and L (3.5μm) is essentially a comparison between flux from the star and flux from the shell and shows a clear link between the shell flux and the occurrence of deep minima in the sense that deep minima occur more frequently around the time that the flux from the shell increases (the variations in L are typically on a time scale of ~1000 days). Feast (1990) notes that although the variation in L might be due to a variation in the mass loss rate from the star, no mechanism for this is known and a more attractive alternative is that a more-or-less constant mass ejection forms a shell somewhere in the region of 20 to 100 stellar radii from the star. At a critical density, dust forms rapidly in clumps, the infra-red flux rises and deep minima become very likely as the dust is blown away by radiation pressure. This activity depletes the shell, so the infra-red flux decreases and deep minima become much less likely. A timescale of ~1000 days for this process seems reasonable, though the model is only qualitative at present.

Finally, Stanford et al (1988) have used spectropolarimetry of R CrB in the 1986 and earlier minima to infer a preferred plane of ejection of dust - a disk rather than a puff - and suggest a non-radial oscillation as a mechanism for ejecting a disk rather than a shell. If this were generally true of RCB stars, however, orientation effects would mean some RCB stars would not show deep minima but would have a strong infra-red excess (Feast 1990) - such stars are not presently known.

To summarise:

1) The "puff" or Loreta-O'Keefe model seems to fit the spectroscopic and photometric observations of deep minima (Feast 1986).

2) The obscuring material is probably carbon in an amorphous or glassy state with grain sizes ~100-500Å.

3) Modelling deep minima in detail is difficult because of geometrical effects (line-of-sight, angular size of puff/star) and the fact that stellar temperature, grain size and size distribution, grain composition, density and the formation and possibly fragmentation rates are all important factors - there are too many free parameters and each deep minimum may therefore appear (and be) unique to the observer.

As with the pulsation phenomenon, it appears that longer baseline data and more detailed observation and modelling of the deep minima are required. It seems clear that better results can be achieved with more extensive coverage (ultraviolet to infra-red, spectroscopy, polarimetry etc.).

Acknowledgements

I would like to thank Mike Feast and Don Pollaco for discussions which greatly helped in the writing of this review and Mike Feast for providing pre-publication results from the SAAO long-term infra-red monitoring programme.

References

Alexander, J.B., Andres, P.J., Catchpole, R.M., Feast, M.W., Lloyd Evans, T., Menzies, J.W., Wisse, P.N.J., & Wisse, M.,1972. MNRAS, 158, 305.

Cottrell, P.L., Lawson, W.A. & Buchhorn, M., 1990. MNRAS, 244, 149.
Drilling, J.S., 1986. I.A.U. Colloquium 87, 9 (review paper).
Drilling, J.S. & Hill, P.W., 1986. I.A.U. Colloquium 87, 499.
Efimov, Yu. S., 1988. Sov. Astr. 65, 512.
Fadeyev, Y.A., 1986. I.A.U. Colloquium 87, 441.
Fadeyev, Y.A., 1988. MNRAS, 233, 65.
Feast, M.W., 1975. I.A.U. Symposium 67, 129.
Feast, M.W., 1986. I.A.U. Colloquium 87, 151 (review paper).
Feast, M.W., 1990. in "Confrontation between stellar pulsation and evolution" (in press) Bologna.
Fernie, J.D., 1982. PASP, 94, 172.
Fernie, J.D., 1989. PASP, 101, 166.
Fernie, J.D., 1990. PASP, (in press).
Gaustad, J.E., Stein, W.A., Forrest, W.J. & Pipher, J.L., 1988. PASP, 100, 388.
Gillett, F.C., Blackman, D.E., Beichmann, C. & Neuegebauer, G., 1986. ApJ, 310, 842.
Glass, I.S., 1978. MNRAS, 185, 23.
Goldsmith, M.J. & Evans, A., 1986. Irish Astr. J., 17, 308.
Goncharova, R.I., Kovalchuk, G.U. & Pugach, A.F., 1983. Astrophysics, 19, 161.
Hecht, J.H., Holm, A.V., Donn, B. & Wu, C.-C., 1984. ApJ, 280, 228.
Holm, A.V. & Doherty, L.R., 1988. ApJ, 328, 726.
Holm, A.V., Hecht, J., Wu, C.-C. & Donn, B., 1987. PASP, 99, 497.
Hunger, K., Schönberner, D. & Rao, N.K., (editors) 1986. 87th I.A.U. Colloquium "Hydrogen-deficient stars and related objects". D. Reidel, Dordrecht.
Jones, K., van Wyk, F., Jeffery, C.S., Marang, F., Shenton, M., Hill, P.W. & Westerhuys, J., 1989. SAAO Circular 13, 39.
Kilkenny, D., 1982. MNRAS, 200, 1019.
Kilkenny, D. & Whittet, D.C.B., 1984. MNRAS, 208, 25.
Kilkenny, D. & Marang, F., 1988. MNRAS, 238, 1P.
Kilkenny, D., Marang, F. & Menzies, J.W., 1988. MNRAS, 233, 209.
Lambert, D.L., 1986. I.A.U. Colloquium 87, 127 (review paper).
Lawson, W.A. & Cottrell, P.L., 1988. MNRAS, 231, 609.
Lawson, W.A. & Cottrell, P.L., 1990. MNRAS, 242, 259.
Lawson, W.A. & Cottrell, P.L., 1989. MNRAS, 240, 689.
Lawson, W.A., Cottrell, P.L., Kilmartin, P.M. & Gilmore, A.C., 1990. MNRAS, (in press).
Marang, F., Kilkenny, D., Menzies, J.W. & Spencer Jones, J.H., 1990. SAAO Circular 14, (in press).
Marraco, H.G. & Milesi, G.E., 1982. AJ, 87, 1775.
McNaught, R.H., 1988. Southern Astronomy, Sep/Oct, p40.
McNaught, R.H. & Dawes, G., 1986. Inf. Bull. var. Stars, 2928.
Menzies, J.W., 1986. I.A.U. Colloquium 87, 207.
Percy, J.R., Carriere, L.E.M. & Fabro, V.A., 1987. AJ, 93, 200.
Pugach, A.F., 1977. Inf. Bull. Var. Stars 1277.
Pugach, A.F., 1984. Sov. Astr., 28, 288.
Raveendran, A.V., Ashoka, B.N. & Rao, N.K., 1986. I.A.U. Colloquium 87, 191.
Rao, N.K. & Nandy, K., 1986. MNRAS, 222, 357.
Rao, N.K. Ashok, N.M. & Kulkarni, P.V., 1980. J. Astrophys. Astr., 1, 1.
Rao, N.K., Giridhar, S. & Ashoka, B.N., 1990. MNRAS, 244, 29.
Saio, H., 1986. I.A.U. Colloquium 87, 425 (review paper).
Schönberner, D., 1977. A&A, 57, 437.

Schönberner, D., 1986. I.A.U. Colloquium 87, 471 (Review paper)
Stanford, S.A., Clayton, G.C., Meade, M.R., Nordsieck, K.H., Whitney, B.A., Murison, M.A., Nook, M.A. & Anderson, C.J., 1988. ApJ, 325, L9.
Walker, H., 1986. I.A.U. Colloquium 87, 407 (review paper)
Weiss, A., 1987. A&A, 185, 178.
Wright, E.L., 1989. ApJ, 346, L89.
Yuin, C., 1948. ApJ, 107, 413.

Discussion

Albrecht: Do not the carbon blobs remain in orbit around the star? This might form a shell later. Some shells from blobs originating at other points should just show up as red dimming only.

Kilkenny: The condensed material, probably glassy or amorphous carbon, will be driven away from the star initially by radiation pressure. The fact that we see an infra-red "excess" means that there is dust around the star but this will probably escape to the interstellar medium eventually.

Kellomäki: I have read that RCB stars are rare but we, however, know a large part of these stars of our Galaxy. Is this true?

Kilkenny: Probably. They should be easy to detect because of the very deep minima (up to 8 or 9 mag). In Table 1 I list about 35 stars so that they must be very rare in space but, because their lifetimes are probably very short, perhaps a few thousand years, a significant fraction of all stars might pass through this stage of evolution.

Samus: We have found that we are able to measure the radial velocity of R CrB with our CORAVEL-type spectrometer and are now trying to measure it every clear night. Can it be of some use for data interpretation and do you recommend to continue?

Kilkenny: Yes, it would be very useful. There is not much published continuous velocity monitoring of RCB stars though I know Dr. R. Griffin has monitored R CrB itself. From my comments on the danger of confusing pulsational and obscurational minima, it would be most useful if photometric data (preferably multi-colour, e.g. UBVRI) could be obtained nearly simultaneously with the radial velocities.

Percy: Are there variations in the pulsational amplitude of stars like RY Sgr which may be correlated with the degree of "activity" (mass loss)?

Kilkenny: It is observed that the pulsation amplitude is sometimes larger as the star emerges from a deep minimum but the reason for this is not known.

Percy: But could the mass loss be connected with pulsation as in the Be stars?

Kilkenny: Yes, it is known that shock waves occur in the pulsational cycle of RY Sgr, for example, as line splitting has been observed at certain phases.

Mattei: We have been informed just 3 days ago by the Central Bureau of Astronomical

Telegrams that UX Ant, at maximum since 1975, has started to fade.

Kilkenny: Good. I was not aware that series of observations of UX Ant existed. It would be useful to follow this star and to publish the results.

Table 1. RCB and Related Stars (Adapted from Drilling and Hill 1986)

	RA (2000) Dec	Vmax	Comments/references
RCB Stars			
XX Cam	04 08 39 +53 21 39	7.3	HdC ?
HV 5637	05 11 32 -67 56 00	14.8	LMC
W Men	05 26 24 -71 11 18	13.9	LMC
HV 12842	05 45 03 -64 24 24	13.7	LMC
SU Tau	05 49 06 +19 04 00	9.7	
UX Ant	10 57 08 -37 23 39	12.2	Kilkenny & Westerhuys Obs.110,90 1990
UW Cen	12 43 17 -54 31 41	9.1	
Y Mus	13 05 48 -65 30 48	10.4	
NSV 6708	14 34 48 -39 33 19	7.5	McNaught & Dawes IBVS 2928
S Aps	15 09 25 -72 03 45	9.9	
R CrB	15 48 34 +28 09 24	5.8	
RT Nor	16 24 19 -59 20 42	10.2	
RZ Nor	16 32 42 -53 17 09	11.0	
V517 Oph	17 15 20 -29 05 26	12.6	Kilkenny (unpublished)
WX CrA	18 08 50 -37 19 46	10.4	
V3795 Sgr	18 13 24 -25 47 24	11.0	
VZ Sgr	18 15 09 -29 42 24	10.2	
RS Tel	18 18 51 -46 32 54	9.8	
GU Sgr	18 24 16 -24 15 29	10.1	
FH Sct	18 45 15 -09 25 49	12.1	Tsessevich Astr.Tsirk. 1169
V CrA	18 47 32 -38 09 31	10.2	
SV Sge	19 08 12 +17 37 42	10.4	
RY Sgr	19 16 33 -33 31 18	6.2	
V482 Cyg	19 59 44 +33 58 30	12.1	Quadruple ? H-rich ?
U Aqr	22 03 20 -16 37 40	11.2	Sr-rich (s-process)
UV Cas	23 02 13 +59 36 42	10.6	
Hot RCB(?) Stars			
DY Cen	13 25 34 -54 14 47	12.5	
V348 Sgr	18 40 20 -22 54 29	11.8	
MV Sgr	18 44 32 -20 57 16	12.7	
HdC Stars			
HD137613	15 27 48 -25 10 11	7.5	
HD148839	16 35 46 -67 07 37	8.3	
HD173409	18 46 27 -31 20 34	9.5	
HD175893	18 58 47 -29 39 17	9.3	
HD182040	19 23 10 -10 42 10	7.0	

Table 2. Some Stars Shown Not to Be RCB Stars

Star	Type	Reference
Z Cir	Me	Feast (1965) IBVS 87
LR Sco	F sg.	Giridhar, Rao & Lambert (1990) Observatory (in press)
SY Hyi	M	Feast (1978) IAU Coll 46, 246
	M5/6	Lawson et al. (1989) J.Astrophys.Astron. 10, 151
V504 Cen	CV	Kilkenny & Lloyd Evans (1989) Observatory 109, 85
AE Cir	Symb.	Kilkenny (1989) Observatory 109, 88
V618 Sgr	Symb.	Kilkenny (1989) Observatory 109, 229
MT Pup	CV	Kilkenny (1989) Observatory 109, 299
V803 Cen	IBWD	O'Donoghue et al. (1987) MNRAS 227, 347
V731 Sco	H-str.	Feast (1975) IAU Symp 67, 129
V1860 Sgr	H,G band	Feast (1975) IAU Symp 67, 129
V605 Aql	PN	Seitter (1989) IAU Symp 131, 315
BG Cep	Be	Pollaco et al (in preparation)
LO Cep	Ae?	" " "
CT Vul	H-str.	" " "
CC Cep	F-G em	" " "
V1405 Cyg	G	" " "

THE HOT R CRB STAR V348 SGR AND THE EVOLUTIONARY STATUS OF THE R CRB STARS

Don Pollacco
Department of Physics and Astronomy
St. Andrews University
Fife KY16 9SS
Scotland

The evolutionary status of the hydrogen-deficient R Coronae Borealis (R CrB) stars has been the source of much controversy over the years. Mixing scenarios have failed to relate the observed photospheric abundances to those found in ordinary stars. Recently two rather exotic scenarios have emerged that address this problem:

The last thermal pulse scenario (Renzini 1979, 1981). Calculations have shown that if a white dwarf suffers a thermal pulse it may be intense enough to re-ignite a helium burning shell sending the star towards the AGB for a second time. This process is accompanied by large scale mixing of the photosphere. Subsequent evolution is via a Schönberner track to the white dwarf configuration.

The merged white dwarf scenario (Webbink 1984). It is thought possible that a double degenerate binary consisting of He and CO white dwarfs may lose sufficient angular momentum by gravitational radiation/material ejection to allow the system to merge within a Hubble time. In these circumstances it is expected that the lighter He dwarf will be smeared around the other producing the observed abundances.

The first clues to the evolutionary status of the R CrB stars may have been found by Herbig (1949, 1968) who while observing the proto-type R CrB in deep minima found evidence of [O II] $\lambda 3727/9$A emission. These lines only occur in low temperature and density material and suggest that R CrB may be surrounded by a low-surface-brightness nebula. IRAS observations of R CrB have revealed the presence of a huge fossil dust shell (Walker 1986) some 8 pc in diameter. Gillett et al. (1986) tried to understand the heating of the shell but concluded that the stellar and interstellar radiation fields are far too feeble to account for the observed shell temperature.

The hot R CrB stars are thought similar to the R CrB stars but have much higher photospheric temperatures (Pollacco 1989). The brightest member of this class, V348 Sgr, is known to be surrounded by a faint nebula. Pollacco et al. (1990a) have studied this object in detail and found that the nebular ionisation and extent require a significantly hotter central body than inferred for this object. They also found indications of a large hydrogen abundance gradient between the star and the nebula (see Fig. 1). UV spectra do not give any indications of another body in

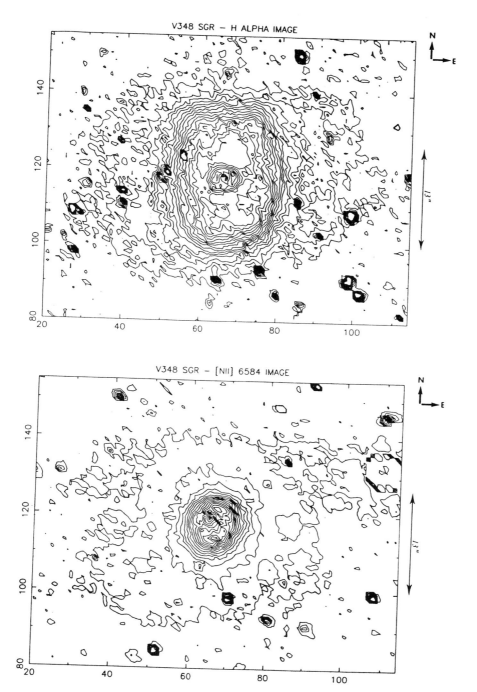

Figure 1: Narrow band Hα and [NII] λ6584Å imaging of the nebula surrounding the hot R CrB star V348 Sgr. The striking morphology difference between the recombination and forbidden line monochromatic images is indicative of a severe hydrogen abundance gradient between the star and nebula, and is in agreement with long-slit spectrographic observations.

the system so they are forced to accept that V348 Sgr must have been hotter in the recent past (the recombination timescale of the nebula is 120-2300 yr). We conclude that V348 Sgr is rapidly evolving towards the R CrB region.

In order to test this scenario Pollacco et al. (1990b) have used ESO's NTT to search for faint nebulae around other R CrB stars. Multiple images were obtained of UW Cen in a very deep minimum that clearly show the star to be surrounded by a faint (and small) nebula of unusual physical appearance (see Fig. 2). The morphology consists of a fainter outer envelope of circular appearance while the central parts are dominated by a pair of reasonably collimated and diagonally opposed "jets". Narrow band imaging and spectra obtained with the NTT suggest that the nebular emission is dominated by a scattering component with no obvious signs of line emission (our detector had little efficiency at $\lambda3720A$).

The nebular structure has proved difficult to understand within the context of the evolutionary scenarios set out above and is more likely related to the characteristic minima observed in these objects. Pollacco et al. (1990b) propose a variation on Feast's (1986, 1990) model for formation of R CrB-minima whereby ejected material which condenses in the circumstellar shell is swept along the jets by the stellar wind and allowed to disperse. This requires the existence of a third pair of jets in the line of sight. With radial velocities of some 200 km s^{-1} as observed in some chromospheric lines during the initial declines to minima for R CrB stars, and assuming that the nebula is expanding at this velocity, an age of some 2500 yr is indicated. This value is not in contradiction with the suspected length of the R CrB evolutionary phase (Schönberner 1977).

Future observations of other R CrB stars in deep minima will allow us to confirm or reject this model and may shed light on the jet collimation mechanism. The detection of extended line emission will lend strong support for Renzini's ideas.

This work is carried out in collaboration with C.N. Tadhunter (RGO), P.W. Hill (St. Andrews), L. Houziaux (Liège), and J. Manfroid (Liège).

References

Feast, M.W. 1986. in: IAU Coll. No. 87, Hydrogen-Deficient Stars and Related Objects, p151, eds. Hunger, K., Schönberner D. and Rao, N. Kameswara. (Reidel, Dordrecht)
Feast, M.W. 1990. in: Confrontation Between Stellar Pulsation and Evolution, Bologna, Italy, (A.S.P. Conference Series), in press
Gillett, F.C., Backman, D.E., Beichman, C. & Neugebauer, G., 1986. ApJ, 310, 842
Herbig, G. 1949. ApJ, 110, 143
Herbig, G. 1968. Mem. Soc. R. Sci. Liège, 17, 353
Pollacco, D.L., 1989. The Evolutionary Status of the Hot R CrB Stars, Ph.D. Thesis, St. Andrews University

Pollacco, D.L., Tadhunter, C.N. & Hill, P.W., 1990a. MNRAS, 245, 204
Pollacco, D.L., Hill, P.W., Houziaux, L. & Manfroid, J., 1990b. Accepted for publication in pink pages of MNRAS.
Renzini, A., 1979. in: Stars and Stellar Systems, p.135, ed. Westerland, B.E. (Reidel, Dordrecht)
Renzini, A., 1981. in: IAU Coll. No 59, Effects of Mass loss on Stellar Evolution, p.319, Eds. Chiosi, C. & Stalio, R. (Reidel, Dordrecht)
Schönberner, D. 1977. A & A, 57, 437
Walker, H.J. 1986. in: IAU Coll. No. 87, Hydrogen-Deficient Stars and Related Objects, p407, eds. Hunger, K., Schönberner D. and Rao, N. Kameswara. (Reidel, Dordrecht)
Webbink, R.F. 1984. ApJ, 277, 355

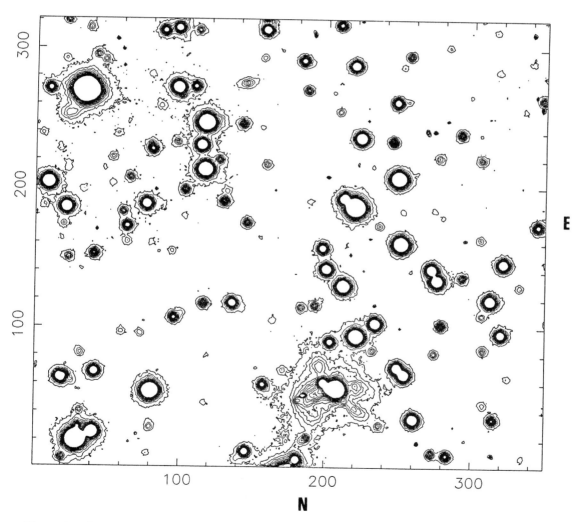

Figure 2: The combined V band image of the R CrB star UW Cen in deep minima. For discussion see text.

CHAPTER 6

RED VARIABLE STARS

OBSERVATIONAL PERSPECTIVES ON RED GIANTS

F.R. Querci and M. Querci
Observatoire Midi-Pyrénées
14 Ave. E. Belin
31400 Toulouse, France

B. Fontaine
Carniol
04150 Banon, France

A. Klotz
Association T 60
14 Ave. E. Belin
31400 Toulouse, France

1. Introduction

In this review we deal with the red giant (and supergiant) stars of the cooler spectral types, that is to say of M, S, and C types. For the historical view and the main physical characteristics of these stars, we refer the reader to previous reviews such as Querci (1986a). Here we focus on the *variability* of these red stars. Strictly speaking, all of them show evidence of variability: variations in the brightness and/or variations in the spectrum (i.e., in line intensity, line radial velocity and line profile). More detailed descriptions of the variability of the red stars may be found in Querci (1986a), Querci (1986b) and Querci & Querci (1989).

The presence of long-term brightness variations (on time scales of years which, particularly, define the primary period of the star) is well established, mainly thanks to amateur astronomers. As the observations become technically more evolved as well as the mathematical tools to reduce them, short-term variations (on time scales of days to months, even of the order of hours) are detected in the red giants.

The main question not yet fully understood is what physics, what phenomena sustain such a variability. Systematic observations for detecting variations on various time scales are urgently needed.

We shall endeavour to summarize the observational features, their present interpretation, the cautions to keep in mind when observing or when reducing the data. Finally some selected observations will be proposed to amateur astronomers.

2. The well-known long-term variability and its interpretation

Observations extended over long time scales led to the classification of the stars in various variability types. Three main types of red *pulsating* variables are defined: the Miras, the semiregulars (or SR types) and the irregulars (or L types). Some red giants belong to the *eruptive* variables such as the RCB (R Cor Bor) stars and the helium stars. The physical intrinsic parameters of the stars play a sensitive role in the type of variability.

Among the red pulsating variables, the *Miras* are the best studied. They follow a global radial stellar pulsation, the modelling of which is due to pioneer studies by P.R. Wood (1979, 1981) or L.A. Willson (1982) and collaborators (Hill & Willson 1979; Fox & Wood 1982; Bowen 1988). Observed and predicted Mira velocities indeed favour that the Miras are pulsating in the fundamental mode for the primary period P_0 (Wood 1990), not excluding other pulsation modes (overtones). The growth or decay with time of the pulsation depends mainly on the variation of the convective energy transport during the pulsation cycle, since in the Miras more than 99 per cent of the flux is carried by convection (Wood 1990). So, these stars are not perfect clocks, and, in fact, even in their prototype, o Ceti, also called Mira, cycle-to-cycle variations are observed.

The pulsation drives acoustic waves which give a shock wave in each pulsation cycle, dissipating energy produced in the lower photosphere. The shock front is produced as the matter forced outward collides with the matter of the previous cycle falling back onto the star. A signature of a shock wave progressing in the atmosphere is the characteristic S-shaped form of the photospheric absorption line radial-velocity curves (e.g. see Fig. 2.1 in Querci 1986b). Another typical property of the Miras is the presence of very intense Balmer emission lines appearing around maximum light (except for the very faint ones). At this time, their profiles are severely mutilated by several superposed sharp absorption lines. With advancing phase, the absorptions disappear and the lines are intense and narrow, showing that the hydrogen-forming layers have progressed above the absorbing layers in conjunction with the progressing shock front, and probing a chromospheric activity (Judge 1989). Generally speaking, comparative line radial velocity curves allow to follow the shock wave running through the stellar atmosphere, thanks to the phase lag between them (e.g., see Fig. 2.6 in Querci 1986b).

The *semiregular* variable light curves are characterized by a form of periodicity hidden by more marked irregular brightness fluctuations.

The SRa present a Mira-like behaviour with, however, a smaller light curve amplitude (≤ 2.5 mag) and strong variations from cycle-to-cycle in the amplitude and the shape of the light curve. An illustration is given in Figure 1.

The SRb have a poorly expressed periodicity as shown in Fig. 2. It may happen that temporary periodic oscillations alternate with slow irregular variations or with a constancy of the brightness such as in RZ UMa (Fig. 3).

In the semiregulars, for which the spectroscopic observations are rare unfortunately, the S-shaped line radial-velocity curves, signature of a shock-pulsation motion, are also recognized, though the curves have a much lower amplitude. Also, a Mira-like phase behavior is noted in hydrogen emission in some stars episodically, for example in the supergiant μ Cep by McLaughlin (1946), in the carbon stars RR Her (SRb) and V Hya (SRa) by Sanford (1950). A stochastic behavior of observed individual features with their highly time-dependent variable profiles is reported: for example, the ultraviolet Fe II emission lines such as in TW Hor (Querci & Querci 1985), the Mg II h and k emission lines such as in α Ori (Dupree et al. 1987) or several metallic lines in this supergiant (Smith et al. 1989).

That the acoustic shock waves are the dominant heating mechanism in the Miras and giant and supergiant non-Miras, is proved by the observational results. Theoretical chromosphere models allow to test the role of acoustic shock waves in the heating (Ulmschneider 1989; Cuntz 1990a,b). Nevertheless, the physical interpretation of transient or highly variable chromospheric features is still unclear. An attractive model is presented by Cuntz (1987) who studies interacting packets of acoustic waves of different periods running through the atmosphere with different speeds, giving a stochastic wave field.

Irregular variables (Lb are giants, Lc are supergiants) present no evident trace of periodicity in their light curves as their name indicates. However, with longer time-series of observations, it happens that they may shift to the semiregular type. A good example is VY Leo for which a long series of uninterrupted observations by satellite demonstrated that this star is an SRa type variable (Maran et al. 1980), whereas the GCVS classified it as an Lb variable.

As detailed in Querci & Querci (1989), the physical mechanisms leading to non-periodic phenomena as observed in the light curves of the radially pulsating red giant variables are still not fully understood.

The presence of various periods in some stars is detected by periodograms. As the astrophysical data are unevenly spaced in time, the prediction of a multiperiodicity requires sophisticated methods to be credible, such as Horne and Baliunas' (1986) treatment applied to α Ori (SRc type) by Karowska (1987): a 1-year period is superimposed on a main cyclic variation of about 5.8-year period (Fig. 4).

An attractive mechanism for the irregular variability in the red variables is proposed by Buchler and Goupil (1988) on the basis of non-linear (chaotic) dynamics. It is able to generate erratic or regularly modulated oscillations.

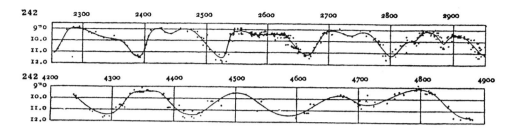

Fig. 1 - Parts of the light curve of S Aq1 (SRa) - Upper: a double maximum appears - Lower: about 4 years after the last observations shown in the upper light curve: the curve oscillates without secondary minima (from Jacchia 1933).

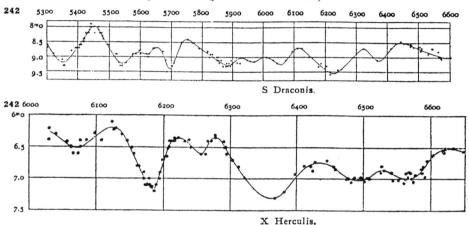

Fig. 2 - Two SRb stars showing their quite individual light curves (from Jacchia, 1933).

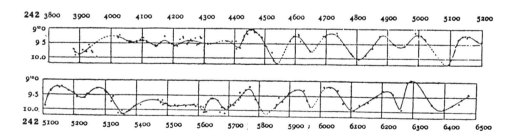

Fig. 3 - Light curve of RZ UMa (SRb): occasionally a near constancy of brightness is evident during about a year (from Jacchia, 1933).

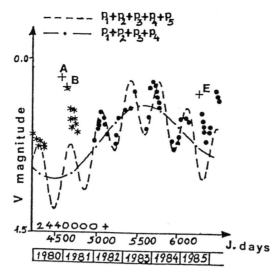

Fig. 4 - The SRc variable, α Ori - Measurements of visual magnitudes by Krisciunas (stars) and Guinan (dots), and synthetic light curves based on superimposition of 4 and 5 sinusoïdal type variations (from Karovska 1987).

3. Short-term variations

3.1 Observational evidence

Short-term variations in brightness and in spectrum are defined as rapid and erratic events. They are on a time scale of the order of a few hours or days to weeks. They have been detected in the last decade in a few "hot" bright supergiants, i.e. in the A and B or F and G spectral type stars (de Jager 1980).

To our knowledge, the first short-term variations involving red giants were discovered in 1973 by Totochava (1975) in RCB stars. During one night, by photometry, she observed a 1.5-hour period in XX Cam and R CrB. Further monitoring of these two stars during many months failed to detect any new palpitations.

Another example is the RCB hypergiant, ρ Cas, in which night-to-night variations were observed by Joshi and Rautela (1978).

As for the pulsating red giants, very few examples were available up until recently since no systematic search for these short-term fluctuations was done. Today the interest is growing in particular among amateur astronomers. Let us summarize the situation.

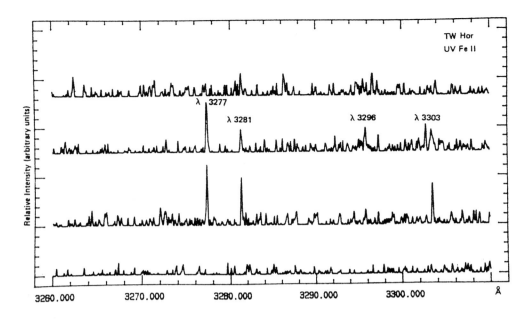

Fig. 5 - *Violet FeII line region in the SRb carbon star, TW Hor, on four consecutive dates in July 1979 (bottom to top: July 13, 14, 15, and 16, respectively) (from Bouchet et al 1983).*

The M8 II type star, R Crt (SRb), was studied by two groups of Brazilian astronomers: Gomes Balboa et al. (1982) found variations at 22 GHz on a time scale of 1 hour; Livi and Bergmann (1982) detected the same variations in the DDO magnitudes and colors in the TiO bands and the Ca I line.

Intensive observations of TW Hor (C3 II; SRb) were done by visible photometry as well as by spectroscopy from the ground (at ESO / La Silla, Chile) and from space with the IUE satellite. Bouchet (1984) notes strong and rapid oscillations in the (U-B) colors when the star is between the phases 0.8 and 0.9. This result seems to be correlated with the night-to-night variations in the Fe II V1 lines found by Bouchet et al. (1983) (Fig. 5). On nine spectra in the IUE / UV range obtained over an interval of 3 years, Querci & Querci (1985) detect variation in intensity of all the emission lines. The most striking features are strong Fe II emissions varying from year-to-year by up to a factor of about 10 (which might be explained by Cuntz' (1987) model quoted above), and the Al II U1 lines varying on a time scale of hours, and from one day to the next.

A systematic monitoring of a sample of SRb type stars and also RCB stars was developed with a photometric twin-telescope technique (Querci et al. 1989). Two SRb M giants, BU Gem and TU CVn, have shown a period of 1.5 hour with a B amplitude of 0.015 mag. However, a contamination of the observations by telluric opacity variations may be possible. This point seems to us so important that we devote paragraph 3.2 to it.

L. Rivas (1989), a GEOS amateur from Valencia, Spain, has monitored 61 Her (M4 III; SRb) in the course of 4 years. During some nights, he observed visual variations of 0.5 magnitude. This seemed so strange to many GEOS amateurs and professionals that they said to him: "Hello Luis, you joke!". To conclude on these variations (which are not definitive), photometric observations were done with the 11-inch twin-telescope at Observatoire de Haute-Provence by the Quercis and with the 70-cm telescope at Jungfraujoch by other members of the GEOS group. Also, spectrographic observations were done at the 60-cm amateur telescope (24" reflector) of the Pic-du-Midi Observatory by one of us (A.K.). Results in July 1989 and February 1990 spectroscopic observations included: strong variations of the metallic absorption lines around Hα with a time scale of 1 to 2 hours and a weak change in the Hα intensity itself. Some months after, i.e. in July and August 1990, no variations appear in the core of the Ca II IR lines.

C. Friedlingstein and J. Vanderbroere (1990), two other GEOS amateurs from Brussels, report variations in the star NSV 3739 CMi (M3 III; SRb) up to 0.3 visual magnitude during some nights in January and April 1990. Photometric and spectroscopic observations are planned to confirm these visual ones.

Theorists have a huge imagination and are able to explain all observations, the good and the bad ones. They invoke: - local changes in amount of material, varying column density and temperatures, changes in the average amplitude of turbulent motions of rising convective cells, sporadic ejection of matter through magnetic-field tubes, etc. Consequently, to help in discriminating between the various interpretations, observers have to do adapted, irrefutable, and repeated observations on many stars.

3.2 Unambiguous detection of the short-term variations and the probable telluric opacity fluctuations

As all photometrists know, some "good" photometric nights are used to test the instrumentation and to deduce the transformation coefficients. Let us emphasize some striking observations done at the Haute-Provence Observatory with a twin telescope (two 11-inch Celestrons) already mentioned above.

On January 2, 1989, the sky was perfectly photometric, and we monitored the binary star HU Tau. The light curve conformed to the curve of an eclipsing binary, without problem. On December 30, again the sky was perfect and one of us (B.F.) monitored some Pleiades (standard stars). To our great surprise, an oscillation appeared in the three filters V, B and U superimposed to the Bouguer line. As no technical reason could explain such a periodicity, there is a strong presumption that the oscillation is due to a variation in the atmospheric extinction (Harvey 1988; Gheonjian et al. 1990). It has a 1.28-hour period and is present in each channel; the amplitude in the V filter is 0.027 mag. The difference of magnitude between Pleiade C and Pleiade D demonstrates that the twin-telescope technique can eliminate the sky fluctuations quasi-totally. Such telluric variations should disturb the observations done by the photometric classic techniques much more than by the twin-telescope technique.

Let us recall that the twin-telescope photometry technique, by monitoring the variable and its comparison simultaneously, was tested to be able to give a measurement accuracy really better than the magnitude variation of the atmospheric extinction on each channel during an observation.

In conclusion, when doing photoelectric photometry, we ought to monitor the atmospheric conditions, for example with a third telescope which follows a check star.

4. Proposed observations for amateurs and conclusion

Two kinds of observations are required. The first involves the technique well known by the AAVSO members; we mean photometry, either by eyes, or by PM or by CCD. The second involves a development of a new technique in the AAVSO: spectroscopy.

Systematic search of various time scale variations (mainly short-term ones) is a new field of research in red giant stars and in many other stars. Previous reviews give a list of stars and spectral features to be monitored (Querci & Querci 1986, 1988). The aim is to disentangle global and local events at the surface of the red stars for a better understanding of their nature.

It is strongly suggested that a large number of amateurs regularly observes a sample of stars selected inside a collaboration with professionals.

References

Becker, S.A. 1987, in "Stellar Pulsation", eds. A.N. Cox, W.M. Sparks, & S.G. Starrfield (Heidelberg:Springer-Verlag), p.16
Bouchet, P. 1984, A&A, 139, 344
Bouchet, P., Querci, M., Querci, F. 1983, The Messenger 31, 7
Bowen, G.H. 1988, ApJ, 329, 299
Buchler, J.R. & Goupil, M.J. 1988, A&A, 190, 137

Cuntz, M. 1987, A&A, 188, L5
Cuntz, M. 1990a, ApJ, 349, 141
Cuntz, M. 1990b, ApJ, 353, 255
Dupree, A.K., Baliunas, S.L., Guinan, E.F., Hartmann, L., Nassiopoulos, G.E. & Sonneborn, G. 1987, ApJ, 317 L85
Fox, M.W. & Wood, P.R. 1982, ApJ, 259, 198
Friedlingstein, C. & Vanderbroere, J. 1990, to be published in GEOS Circulars
Gheonjian, L.A., Klepikov, V.Yu. & Stepanov, A.I. 1990, preprint n°4, Inst. Terrestrial Magnetism, Ionosphere and Radio-wave Propagation, Moscow
Gomes-Balboa, A., Lépine, J.R.D. & Pires, N. 1982, private communication
Harvey, J.W. 1988, in "Advances in Helio- and Asteroseismology", eds. J.Christensen-Dalsgaard and S. Frandsen (Dordrecht:Reidel), p. 497
Hassenstein, W. 1938, Pub. Astr. Obs. Potsdam 29, Part 1
Hill, S.J. & Willson, L.A. 1979, ApJ, 229, 1029
Horne, J.H. & Baliunas, S.L. 1986, ApJ, 302, 757
Jacchia, L. 1933, Le Stelle Variabili, Pub. Osservatorio Astronomico, Univ. di Bologna, Vol. II, n°14
de Jager, C. 1980, The Brightest Stars, Dordrecht:Reidel
Joshi, S.C. & Rautela, B.S. 1978, MNRAS, 183, 55
Judge, P.G. 1989, in "Evolution of Peculiar Red Giant Stars", eds. H.R. Johnson and B. Zuckerman (Cambridge Univ. Press), p. 303
Karowska, M. 1987, in "Stellar Pulsation", eds. A.N. Cox, W.M. Sparks, & S.G. Starrfield (Heidelberg: Springer-Verlag), p. 260
Livi, S.H.B. & Bergmann, T.S. 1982, AJ, 87, 1783
Maran, S.P., Michalitsianos, A.G., Heinsheimer, T.F. & Stecker, T.L. 1980, in "Current Problems in Stellar Pulsation Instabilities", eds. D. Fischel et al., NASA TM-80625, p. 629
McLaughlin, D.B. 1946, ApJ, 103, 35
Querci, M. & Querci, F.R. 1985, A&A, 147, 121
Querci, M. & Querci, F.R. 1989, in "Evolution of Peculiar Red Giant Stars", eds. H.R. Johnson & B. Zuckerman (Cambridge Univ. Press), p. 303
Querci, F.R. & Querci, M. 1986, in "Automatic Photoelectric Telescopes", eds. D.S. Hall, R.M. Genet & B.L. Thurston (Fairborn Press), P. 156
Querci, F.R. & Querci, M. 1988, in "New Directions in Spectrophotometry", eds. A.B. Davis-Philip, D.S. Hayes & S.J. Adelman (L. Davis Press), p. 59
Querci, F.R., Querci, M., Grégory, Cl. & Fontaine, B. 1989, in "Remote Access Automatic Telescopes", eds. D.S. Hayes & R.M. Genet (Fairborn Press), p.35
Querci, F.R. 1986a, in "Non-thermal Phenomena in Stellar Atmospheres, The M, S, and C Type Stars", eds. H.R. Johnson & F.R. Querci, NASA SP-492, p. 1
Querci, M. 1986b, in "Non-thermal Phenomena in Stellar Atmospheres. The M, S, and C. Type Stars", eds. H.R. Johnson & F.R. Querci, NASA SP-492, p. 113
Rivas, L. 1989, Aster 110, 18
Sanford, R.F. 1950, ApJ, 111 270
Smith, M.A., Patten, B.M. & Goldberg, L. 1989, AJ, 98, 2233
Smith, M.A. 1976, Journal AAVSO 5, 67

Totochava, A.G. 1975, in "Variable Stars and Stellar Evolution", eds. V.E. Sherwood & L. Plaut (Dordrecht:Reidel), p. 161
Ulmschneider, P. 1989, A&A, 222, 171
Willson, L.A. 1982, in "Pulsations in Classical and Cataclysmic Variables", eds. J.P. Cox & C.J. Hansen (Boulder: JILA), p. 284
Wood, P.R. 1979, ApJ, 227, 220
Wood, P.R. 1981, in "Physical Processes in Red Giants", eds. I. Iben & A. Renzini (Dordrecht: Reidel), p. 205
Wood, P.R. 1990, in "From Miras to Planetary Nebulae: Which Path for Stellar Evolution", eds. M.O. Mennessier & O. Omont (Editions Frontières), p. 67

Discussion

Williams: You seemed to assign some considerable significance to a relationship between the acoustic velocity in a star and a variable you called the "Chaotic Attractor". Could you please describe the chaotic attractor in more detail and explain its physical significance.

Querci: Radial global pulsation drives acoustic waves. Several pulsation modes exist and interact. A star may be periodic or multiperiodic. If not, it is aperiodic. Two kinds of aperiodicity are known: randomness and chaos. Chaos is when the motion is deterministic (e.g., see Perdang, J. 1985, in "Chaos in Astrophysics", eds. J.R. Buchler, J. Perdang & E.A. Spiegel, Dordrecht: Reidel, P. 11).

Sareyan: (1) The photometer used at Haute Provence Observatory was actually made by Norman Walker and not in our Nice group.

(2) In favour of the twin telescope technique, I must add that the Haute Provence Observatory is not considered as being a photometric site (altitude ~ 600 m, humidity, ---).

Querci: (1) The photometers in the twin-telescope used at Haute-Provence Observatory were built by Bruno Fontaine. Only the optical heads situated behind each telescope were bought from Norman Walker.

(2) Yes, you are right. It is the main reason why we built a twin-telescope which overcomes adverse sky conditions quasi-totally as we comment in the text.

THE ANALYSIS OF OBSERVATIONS OF MIRA STARS

John E. Isles
British Astronomical Association
P.O. Box 6322, Limassol, Cyprus

Introduction

This paper summarizes work in progress that is appearing in a series of reports in the Journal of the BAA (Isles & Saw 1987 [hereafter Paper I], 1989a [Paper II], 1989b, 1989c; Saw & Isles 1990). These discuss visual observations of Mira stars made by members of the BAA VSS. The data are independent of those available from other sources, for example AAVSO. Although the BAA VSS was formed in 1890, it was not until the appointment of Colonel E.E. Markwick as Director in 1900 that a fixed programme of stars to be observed was introduced. Since then, some 2 million observations have been made of several hundred variable stars.

During the first half of the present century, the main emphasis in the work of the BAA VSS was on the observation of Mira stars. More recently there has been increasing concentration on other classes, particularly the eruptive and cataclysmic variables. In order to make room for more of these in the programme, about 30 Mira stars were dropped from our working list in 1974. So far, it is only for these Mira stars, listed in Table 1, that the observations have been analysed in detail.

Results of the Analysis

(a) Classification of Light Curves

Typical light curves for one cycle of each star have been drawn by plotting the three rises and three falls of well-observed cycles that had rise and fall times and maximum and minimum magnitudes nearest to the mean values observed by the BAA VSS. Some examples of these curves are shown in Figure 1.

A qualitative classification of the light curves of long-period variables was given by Ludendorff (1928); see Hoffmeister et al. (1985) for a convenient account of it. For most stars there is quite good agreement between Ludendorff's classification and our own, but there are some stars (for example S Boo, S Cep, T Her) where the BAA data are clearly not consistent with Ludendorff's classification. Our results are, however, generally in close agreement with the mean light curves derived from AAVSO data by Campbell (1955). It is not clear how far the qualitative classification of light curves can help us towards an understanding of what is happening in Mira stars, but there is certainly a statistical relationship

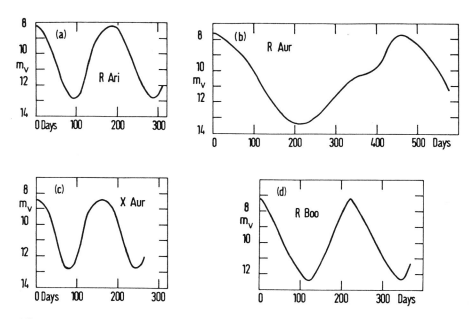

Figure 1. Typical light curves of (a) R Ari; (b) R Aur; (c) X Aur; (d) R Boo

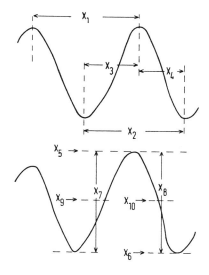

Figure 2. Imaginary light curves showing (above) the time quantities X_1 to X_4 and (below) the magnitude quantities X_5 to X_{10}

between the shape of the light curve and the period, as has been shown by Campbell (1925) and by Tuchman (1984).

(b) Summary Statistics

Table 1 lists for each star the mean period; the mean time taken to rise from minimum to maximum brightness, expressed as a fraction of the mean period; the extreme range in magnitude; and the mean range. These data were derived from maxima and minima as observed by the BAA VSS during the years indicated in the final column. The dates and magnitudes of maxima and minima were found by Pogson's method of bisected chords, and were those on which the star was judged to be actually brightest or faintest. This procedure differs from that used by the AAVSO, in which a mean curve is fitted to the data. The AAVSO method has certain advantages but, for the statistical analyses we wished to make, we thought it was important that we should not risk imposing a spurious regularity on the data.

We have compared the results in Table 1 with corresponding figures given in the General Catalogue of Variable Stars (GCVS, Kholopov, 1985-7). The present, Fourth Edition of the GCVS does not give the mean ranges of Mira stars; these were last given in the Third Supplement to the Third Edition (Kukarkin et al. 1976). For objects like these whose magnitude ranges are not precisely bounded, the mean range is a more useful quantity to know than the extreme range, which is likely to expand the longer the star is observed. The GCVS does not state the source of its numerical data for any of those stars, though many of them are presumably derived from AAVSO results. Although our figures generally show good agreement with the GCVS, we found several discrepancies that could not be accounted for by differences between AAVSO and BAA comparison star sequences.

(c) Correlations

From the observed dates and magnitudes of maxima and minima, ten quantities of interest were derived for each cycle, as follows (see also Figure 2):

X_1 Cycle length (max. to max.)
X_2 Cycle length (min. to min.)
X_3 Rise time
X_4 Fall time
X_5 Maximum magnitude
X_6 Minimum magnitude
X_7 Amplitude (rise)
X_8 Amplitude (fall)
X_9 Mid-range (rise)
X_{10} Mid-range (fall)

Table 1. Mira Stars Covered By BAA Studies To Date

Star	Period d	(M-m)/P	Extreme range m m	Mean range m m	Years
R Ari	186.92	0.47	7.6 - 13.7	8.2 - 12.8	1903-74
R Aur	457.76	0.52	6.6 - 13.8	7.7 - 13.4	1904-74
X Aur	163.90	0.49	7.9 - 13.5	8.4 - 12.8	1910-74
R Boo	223.51	0.46	6.3 - 12.9	7.2 - 12.3	1906-74
S Boo	271.08	0.48	8.0 - 13.8	8.4 - 13.3	1921-74
R Cam	270.01	0.48	7.9 - 14.5	8.4 - 13.4	1920-69
R Cas	429.92	0.41	5.2 - 13.5	6.4 - 12.6	1901-74
W Cas	405.99	0.48	8.2 - 12.3	8.8 - 11.9	1920-74
S Cep	486.93	0.54	7.7 - 12.4	8.5 - 11.2	1923-74
T Cep	388.20	0.53	5.6 - 11.0	6.0 - 10.2	1901-74
U Cyg	464.99	0.48	6.8 - 12.3	7.6 - 11.1	1908-74
S Del	278.15	0.48	8.6 - 12.4	8.7 - 11.9	1964-74
R Dra	245.34	0.43	6.9 - 12.9	7.5 - 12.3	1906-74
R Gem	369.82	0.35	6.0 - 13.8	7.0 - 13.2	1907-74
S Her	308.27	0.48	6.5 - 13.7	7.4 - 13.1	1904-71
T Her	164.93	0.46	6.7 - 13.8	7.9 - 13.0	1904-74
U Her	405.28	0.41	6.8 - 13.2	7.7 - 12.6	1920-74
R Leo	312.62	0.45	5.2 - 10.6	5.8 - 10.1	1905-74
R Lyn	378.47	0.42	7.2 - 14.0	7.8 - 13.4	1920-74
W Lyr	196.33	0.48	7.4 - 13.0	7.9 - 12.2	1914-74
RY Oph	150.47	0.47	7.4 - 14.4	8.2 - 13.4	1917-71
R Peg	378.65	0.43	7.0 - 13.7	7.8 - 12.9	1905-74
X Peg	200.87	0.49	8.8 - 14.5	9.4 - 13.6	1914-72
R Per	209.93	0.48	8.3 - 14.8	8.8 - 13.8	1922-74
V Tau	169.86	0.48	8.6 - 14.7	9.2 - 13.6	1922-72
R Tri	267.46	0.46	5.6 - 12.4	6.4 - 11.8	1963-74
R UMa	301.56	0.38	6.7 - 13.7	7.5 - 13.0	1904-74
S UMa	225.94	0.48	7.2 - 12.3	7.8 - 11.6	1904-74
S UMi	327.28	0.50	7.9 - 12.9	8.3 - 12.0	1911-72
S Vir	377.30	0.44	5.8 - 13.1	6.9 - 12.6	1906-74
R Vul	136.67	0.48	7.3 - 14.3	8.0 - 12.6	1911-72

We investigated whether any of these ten quantities, together with the corresponding ten quantities for the following cycle which we called X_{11} to X_{20}, were correlated with one another. We found that the correlations were highly significant in a large number of cases, showing that Mira stars do not produce long or short cycles, bright or faint maxima, and so on, completely at random.

The results are difficult to summarize briefly, but we did find one thing. When the same two quantities were correlated in more than one star, we found that the correlation coefficients most often had the same sign. This tends to confirm the family resemblance of Mira stars. It could, indeed, be the case, at least for some of the correlations considered, that they are actually present in the light curves of every Mira star; but as we usually have fewer than 100 observed cycles for each individual star, it could be just a matter of chance whether a correlation turns out to be statistically significant in that star's data.

One correlation we found to be strong in several stars is that between X_1 and X_5: a long cycle (max. to max.) tends to close with a faint maximum. This is consistent with a model in which a shock wave carries energy up to the star's surface, the speed of the shock wave depending on the amount of energy it carries.

Harrington (1965) investigated this same correlation in 165 stars, using 29 years of AAVSO data as given by Campbell (1955). He found that only 41% of the stars showed such a correlation at the 5% significance level, but the distribution of correlation coefficients for all 165 stars was consistent with statistical fluctuations about the mean. Also, there was no discernible difference between the 'significant' stars and the rest, whether one considered period, spectral type or space velocity. In the BAA data for 29 stars observed for the longer average interval of 63 years, this correlation is significant at the 5% level for as many as 17 stars (59%). In long enough runs of data, perhaps every Mira star would show this correlation.

This particular correlation was between a time quantity (cycle length) and a magnitude quantity (maximum brightness). But most of the correlations we found to be significant related times to times, or magnitudes to magnitudes. Occasionally, as in R Ari, there was a tendency for the shape of the light curve to alternate from cycle to cycle. Other stars showed slower, systematic changes in light-curve shape over the course of several or many cycles. R Lyn and W Lyr had gradually varying amplitude, while S Cep and U Cyg had gradually varying mean magnitude. The changes in amplitude and mean magnitude for these stars did not show any evident periodicity.

(d) Period Change

One of the most tricky questions concerning Mira stars is whether their periods are genuinely variable. The answer to this depends on what you mean by the star's period. There are certainly real changes in cycle length from one cycle to the next. But suppose there is some mechanism within the star that acts as though it draws these cycles of various lengths at random from a statistical distribution that has some particular mean value. The question we have tried to answer is whether that mean value, which is what we call the period, is fixed or gradually changing.

I admit that in reality the variations of Mira stars must be deterministic, and we should strictly reject the idea of cycle lengths being generated by a purely random process. Deterministic processes can nevertheless give rise to apparently random behaviour, and so I believe it is informative to consider whether the observed cycle lengths of Mira stars can be distinguished statistically from random series.

We have used three different tests to check this, with results as given in Table 2 (where the short runs of data for S Del and R Tri are omitted). The methods used are discussed further in the poster paper given at this meeting by Chris Lloyd. The results quoted here are those derived from analyses of dates of maxima, as these are generally better observed than minima.

(d) (1) Method of Sterne and Campbell (1936)

This is a chi-square test on a contingency table in which cycles are formed into groups of about 10 and classified according to whether they are longer or shorter than the overall mean. It is capable in principle of detecting any type of gradual period change, though the grouping of data involves discarding relevant information, and this must reduce its sensitivity. The test also suffers from appalling instability in the short term, as has been pointed out by Lloyd (1989).

According to this test, the period of one star, R Aur, is significantly variable at the 1% level according to BAA data, while those of three stars (R Cam, U Cyg and S Vir) are significantly variable at the 5% level. (Note that, with 29 stars being tested, we should expect to find on average 0.29 stars with apparently significant change at the 1% level, and another 1.16 at the 5% level, even if none of the 29 stars have genuinely variable periods.)

(d) (2) Span Test (Paper I)

This test involves first calculating the O-C value for each cycle. This is the difference between the observed (O) dates of maximum, and those calculated (C) on the basis of a constant cycle length. Figure 3 shows the way O-C varies for three of the stars examined.

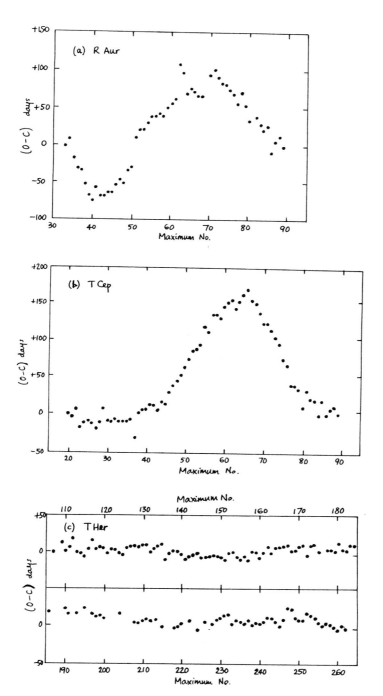

Figure 3. O-C diagrams for (a) R Aur; (b) T Cep; (c) T Her

Table 2. Comparison of results of tests for period change

Star	No. of cycles	Sterne and Campbell Chi^2	d.f.	P	Span test	Lag-2 test P
R Ari	90	4.82	8	0.777		0.217
R Aur	52	16.29	4	0.003**		0.494
X Aur	114	8.83	10	0.548		0.997
R Boo	91	8.08	8	0.885		0.979
S Boo	63	6.34	5	0.275		0.407
R Cam	58	13.00	4	0.011*		0.622
R Cas	62	6.89	5	0.229		0.187
W Cas	43	3.74	3	0.291		0.523
S Cep	32	0.42	2	0.811		0.702
T Cep	69	6.90	6	0.330	**	0.022*
U Cyg	50	10.74	4	0.030*		0.107
R Dra	96	4.25	8	0.834		0.662
R Gem	48	2.00	3	0.572		0.102
S Her	62	9.62	5	0.868	*	0.735
T Her	118	13.12	10	0.217		0.99988
U Her	34	1.06	2	0.589	*	0.104
R Leo	56	8.21	4	0.084	*	0.032*
R Lyn	48	1.59	3	0.662		0.671
W Lyr	97	5.40	8	0.714		0.988
RY Oph	77	4.26	6	0.641		0.932
R Peg	55	3.94	4	0.414		0.247
X Peg	70	2.98	6	0.811		0.260
R Per	59	6.40	5	0.269		0.481
V Tau	46	3.25	3	0.355		0.823
R UMa	78	9.44	6	0.136		0.600
S UMa	109	4.19	10	0.938		0.741
S UMi	66	10.52	5	0.062		0.045*
S Vir	47	8.35	3	0.039*	*	0.108
R Vul	114	6.50	10	0.772		0.736

The symbols * and ** respectively denote significance at the five and one per cent levels.

In the span test, the largest of all the O-C values for a particular star is compared with the distribution of largest O-C values that would be expected for a star with the same standard deviation in cycle length, and without any inbuilt tendency for cycles longer or shorter than the mean to come in any particular order. The test is most sensitive to a single, sudden period change occurring in the middle of the run of data. It could miss altogether a case in which a star's period changes several times, with increases and decreases alternating in such a way that the O-C value never becomes very large.

This test gives a result significant at the 1% level for one star (T Cep), and at the 5% for four (S Her, U Her, R Leo and S Vir).

(d) (3) Lag-2 Correlation Test (Paper II)

In the absence of other effects, a change in the period of a star should give rise to a positive correlation in the lengths of consecutive cycles. However, accidental variations in the date of maximum, whether they be due to observational error or secondary variations in the star's light curve, would give rise to a negative correlation, which could swamp any positive correlation due to period change. Our third test therefore looks not at the correlation between the lengths of cycles that are immediately consecutive (lag 1) but at those separated by one intervening cycle (lag 2).

We have found by this method only three period changes significant at the 5% level (T Cep, R Leo and S UMi). These are stars with small P values in the final column of Table 2. This test also threw up several stars with unexpectedly *large P* values. P is the probability of obtaining (by chance, in the absence of any real period change) a correlation coefficient as high as, or higher than, the value actually obtained. A small enough P value therefore indicates a significant positive correlation, while a large value indicates a negative correlation.

Two stars (X Aur and T Her) had negative correlations that (had we originally set out to test for such a thing) we would have considered significant at the 1% level, and another two (R Boo and W Lyr) at the 5% level. These stars show short-term oscillations in cycle length about the mean value, and we speculate that this may represent phase jitter due to an imperfect coupling between the stars' surface layers (which are all we actually see) and the mechanism within the star driving the variations. If the surface layers, for whatever reason, undergo one or two pulsations longer than average (for example), they then have to produce one or two shorter than average to catch up with the driving mechanism.

Our original hope that method (2) or (3) would prove decisively superior to the traditional method (1) has been disappointed, but the investigation has been informative. Each test yielded between three and five stars with period change apparently significant at the 5% level or better. This number is much higher than the expectation, in the absence of period changes, of 1.45 stars. Thus, there should be at least two or three stars in the sample with variable periods. Yet for not one star did all three tests agree in giving a positive result.

For 20 stars, all three tests gave a negative result, including seven stars (W Lyr, RY Oph, R Peg, X Peg, V Tau, S UMa, R Vul) which the GCVS states to have variable periods. The results suggest that many of the supposed cases of period change in Mira stars are ambiguous, if indeed they are based on any real evidence. Apart from exceptional cases of gross period change such as R Hya, careful statistical tests using long runs of data are needed to establish the reality of any changes.

Recommendations

(1) The above analyses should be extended to include the 34 Mira stars now on the programme of the BAA VSS, some of which have been observed for almost a century, and other Miras for which as long run of data exists.

(2) Observers should continue to follow the Mira stars, *especially* those that are already thought of as well studied, since long runs of data are much more useful than short ones.

(3) Variable-star groups should combine their data to ensure that the runs of data available for analysis are both as complete and as long as possible. This means pooling historical data as well as current observations.

(4) Guidance would be very welcome from theoreticians, especially perhaps those investigating the behaviour of non-linear ('chaotic') systems, as to the way in which analyses of long runs of observations might be most usefully presented.

Acknowledgements

The results reported in this paper are derived from 237,000 observations made by several hundred members of the BAA VSS since 1901. The analysis of these observations in the present form was initiated and guided by Doug Saw, Director of the BAA VSS from 1980 to 1987, whose death in 1990 March was a great loss to astronomy. Herbert Joy assisted greatly in arranging the unpublished observations for analysis. The writer is very grateful to the AAVSO for a grant enabling him to attend this meeting.

References

Campbell, L., 1925, Harvard Repr., No. 21.
Campbell, L., 1955, Studies of Long Period Variable Stars, Cambridge, Mass.
Harrington, J.P., 1965, Astron. J., 70, 569.
Hoffmeister, C., et al., 1985, Variable Stars, Berlin.
Isles, J.E. & Saw, D.R.B., 1987, JBAA., 97, 106.
Isles, J.E. & Saw, D.R.B., 1989a, JBAA., 99, 121.
Isles, J.E. & Saw, D.R.B., 1989b, JBAA., 99, 165.
Isles, J.E. & Saw, D.R.B., 1989c, JBAA., 99, 275.
Kholopov, P.N. (ed.), 1985-7, General Catalogue of Variable Stars, Fourth Edition, I-III, Moscow.
Kukarkin, B.V., et al., 1976, Third Supplement to the General Catalogue of Variable Stars, Moscow.
Lloyd, C., 1989, The Observatory, 109, 146.
Ludendorff, H., 1928, Handbuch der Astrophysik, 6, 99.
Saw, D.R.B. & Isles, J.E., 1990, JBAA., submitted.
Sterne, T.E. & Campbell, L., 1936, Harvard Ann., 105, 459.
Tuchman, Y., 1984, MNRAS., 208, 215.

DETECTING PERIOD CHANGES IN MIRA VARIABLES

C. Lloyd
Astrophysics Division
Rutherford Appleton Laboratory
Chilton, Didcot OX11 OQX, UK

Introduction

Period changes in Miras have been recognized for many years but their origin is still not clear and the proportion of stars affected is a controversial subject. Although Miras are rapidly evolving stars the effects are too slow to be seen. Some observed period changes may be caused by helium shell flashes but period reversals and continuous changes on much shorter time scale are also seen. These may result from chaotic or more predictable behaviour of the pulsation mechanism but need observational clarification.

The light curves of Miras are unstable and so each period is different to the next. In most cases the individual periods vary randomly about a constant mean period. When the times of maximum light are plotted in an O-C diagram the variation of the individual periods causes the points to wander about. These wanderings may mimic real changes, either a constant increase or decrease in period or changes between different constant periods, even though the mean period may be constant (see figure 1). Visual inspection of the O-C diagram is not enough. The reality of apparent period changes has to be estimated from a statistical test of a large sample of stars.

The tests

What the test has to show is whether or not some characteristic of the distribution of periods, O-C residuals or some other observed property of the light variations is purely random. This is not trivial because the large random variations and observational error conspire to mask any real changes in the mean period. Also a test is usually sensitive to only one property of the distribution. The traditional test is due to Sterne and has been used extensively. It measures the probability of getting the observed distribution of periods about the mean for groups of about 10 periods. The span test measures the probability of getting the largest observed residual in the O-C diagram by chance. The lag-2 test measures the degree of correlation between alternate periods. If the periods are random then there should be no correlation. These are described in greater detail in Isles & Saw (1989) and Lloyd (1991a). The Q statistic (Lloyd 1991b) measures the probability of getting the observed total range in the O-C diagram by chance.

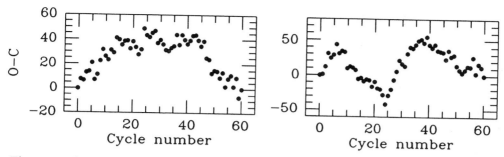

Figure 1: Simulated O-C diagrams of a single constant period and purely random errors which mimic (left) a decreasing period and (right) abrupt period changes.

Figure 2 shows the O-C diagrams and the results of applying the tests to the times of maximum light of R Aurigae and T Herculis. Each test statistic has been calculated for all the data up to each cycle number and this is simply a device for comparing the behaviour of the tests with changes in the O-C diagram. The O-C diagram for R Aurigae (Lloyd 1989) suggests that the period changes between different constant values. But are these period changes real? Sterne's test shows considerable scatter and although most points lie in the critical region towards the end the test becomes confused and gives inconsistent results. The span test suggests that the first period change is significant at the 1% level but it is not convinced by the second or third change. What does this mean? The results from the lag-2 test show a clear and sustained run into the critical region finally reaching a significance of 2%. The period changes look real. And finally, the Q statistic quickly passes through the critical region into the realm of statistical nirvana finally reaching a significance of 0.2%. The period changes look very real. Already from this one example it is clear that the tests behave differently and give different results.

T Herculis is rather different. There is very little activity in the O-C diagram and no suggestion of any period changes. Sterne's test finds nothing usual. In contrast the span test and the Q statistic suggest that the variation in the O-C diagram is *too small* to be random and the lag-2 test reveals a marked negative correlation between alternate periods. These results do not suggest that the mean period is changing, but that the individual periods are not randomly distributed. There is probably some cyclical behaviour on a time scale of a few periods. Here again are contradictory results because (in this case) Sterne's test is not sensitive to the non-randomness in the data.

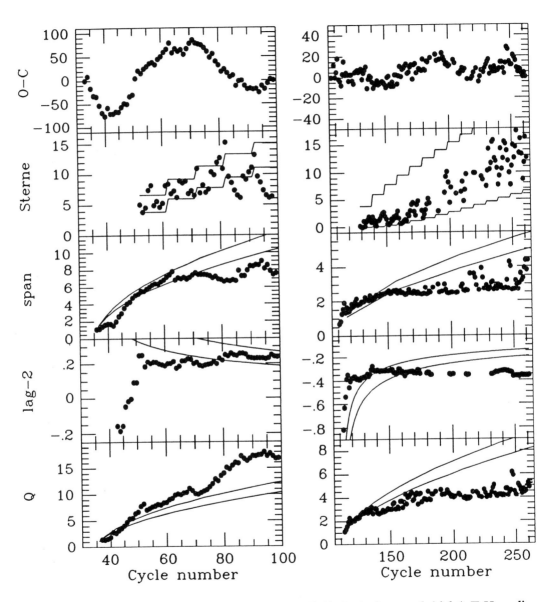

Figure 2: O-C diagram and test statistics for (left) R Aurigae and (right) T Herculis. For R Aurigae the lines show the significance levels of 1% (upper) and 5% for each test. For T Herculis the lines show 5% (upper) and 95% for Sterne's test and 95% (upper) and 99% for the other tests.

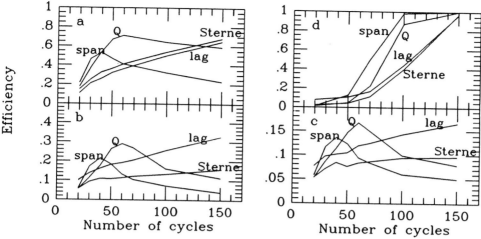

Figure 3(a-c): Efficiency of the tests at detecting period reversals. The period changes by 6 days about every 25 cycles, (a); with no error on the time of maximum light $\sigma_t=0$, error on the period $\sigma_p=6$ days, (b); introduce error on the time of maximum light $\sigma_t=0, \sigma_p=6$, (c); increase error on the period $\sigma_p=10, \sigma_t=6$. Figure 3d: Efficiency at detecting a continuous period change of 6 days over 60 cycles with $\sigma_t=\sigma_p=6$

Efficiency

The different behaviour of the tests is also dramatically illustrated in figure 3 which shows their efficiency at detecting period reversals (like R Aurigae) and continuous period changes in simulated data. The span test is very efficient at detecting one period change and similarly a continuously changing period but is very poor in the face of several period reversals. The Q statistic is the most efficient at detecting period reversals in small samples and nearly as good as the span test for continuous changes. The lag-2 test is the most efficient at detecting period reversals in large samples but for smaller samples and continuous changes it does not work so well. Finally Sterne's test is not very efficient and is also sensitive to the bin structure used. It is also clear the size of the observational error and the intrinsic scatter of the individual periods has a profound effect on the ability of all the tests to detect real period changes. Gaps in the data will also reduce the efficiency of the test; some more than others.

Conclusions

It has been shown that the tests react differently and have very different efficiencies depending on the type of period change and the number of observations. No one test provides all the answers but it is unrealistic to consider samples of stars rather than individuals and the most practical approach is probably to accept period changes if the *appropriate* significance level is reached in at least one test out of a few applied uniformly.

References

Isles, J.E. & Saw, D.R.B., JBAA, 99, 121 (1989)
Lloyd, C., JBAA, (in press)
Lloyd, C., A&A (submitted)
Lloyd, C., Observatory 109, 146 (1989)

PREDICTING THE BEHAVIOUR OF VARIABLE STARS

Marie-Odile Mennessier
USTL Astronomie
F 34060 Montpellier Cedex 1
France

1. Introduction

Variable stars play a very important role in many fields of astronomy, including stellar and galactic evolution, and physics of interstellar and circumstellar media. Moreover, Cepheids are prominent in the calibration of the extragalactic distance scale. Some variable stars are radio sources which will be used in the linking of the satellite reference system to extragalactic objects (Perryman et al 1989).

Due to their importance, 7000 variable stars were proposed as HIPPARCOS observing targets; all could not be observed due to observational constraints. The choice of the ones suitable for observation was made using simulations in the same way as for non-variable stars, but the mean magnitude assigned to each variable star was chosen in view of the shape of the light curve.

2. Problems Related to the Observations of Variable Stars

For optimum use of the available observing time, which is allocated by the star observation strategy implemented in the satellite on-board computer, a good estimate of the magnitude of each star at each observation epoch is needed. Thus, specific problems related to the observations of variable stars arose: if we adopted a unique magnitude, this led to a loss of precision when the star was fainter than the adopted magnitude and a loss of time when the star was brighter (Figure 1). For a good compromise, it was deduced that a luminosity ephemeris was necessary for the observation of "regular" variable stars with an amplitude larger than about 2.5 magnitudes (Mennessier and Baglin 1988). There are 245 such stars and they are mainly long period variables.

3. Predictions

Any predictive method must follow three steps: analysis of past behaviour from a learning sample, check of learning on a test sample, and application of learning to predictions.

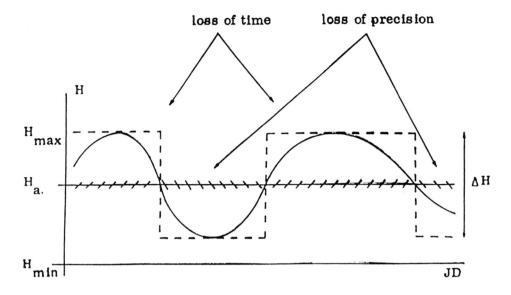

Figure 1. Optimizing the Available Observing Time

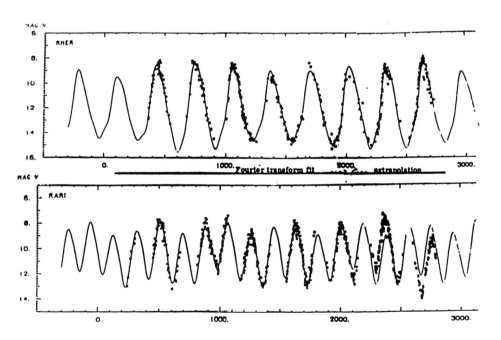

Figure 2. Line: Fourier development from observations during the first 2000 days, extrapolated to the next 1000 days. Dots: observed magnitudes from AFOEV data

3.1. Analysis Methods

Light curves of long period variable stars show differences from one cycle to another that must be taken into account. The light curves are a pseudo-periodic chronological series with missing data (Figure 2) (Mennessier 1989).

A preliminary study applied to Association Francaise des Observateurs d'Etoiles (AFOEV) 7-year data showed that classical methods were not well suited to the problem (Figure 2). Only 25% of the light curves could be correctly predicted by such methods (Mennessier et al. 1989a).

We performed two analyses and used them as complementary (Figure 3) to the classical methods. The first approach was derived from Fourier analysis. The principal difficulty was to separate the real frequencies from the spurious ones introduced into the analysis by the missing data. By applying this method to AAVSO 20-year individual observations, we determined a preliminary power spectrum of the light curve.

The second approach (Mennessier and Diday 1989; Quinqueton 1989; Mennessier et al. 1989b) was a learning method based on artificial intelligence procedures. The basic idea was to learn rules linking the characteristics of two, three, or even more consecutive cycles to one characteristic of a later cycle. We learned rules using AAVSO 75-year dates and magnitudes of extrema (these data go back further than the 20-year ones do).

3.2. Predictions

According to the results of the checks for each considered star, the predictions were deduced from a Fourier development and/or from rules.

R Fornacis is a good example (Mennessier et el. 1990). The rules allowed us to suspect a long-term variation that could not be found from a Fourier analysis of data observed for only 20 years. By adding a low frequency to the preliminary power spectrum as the starting points of the non-linear fit, we improved the fit (Figure 4), and we may consider the resulting Fourier development as characterizing the true behaviour of the light curve.

4. Results of Predictions

For the HIPPARCOS mission, the predictions prepared at Montpellier every two months are checked using both visual and satellite observations.
The visual observations are made by amateur astronomers and compiled by the AAVSO. Very rapid communications are used for allowing J. Mattei, Director of the AAVSO, to check the predictions against the most recent possible observations.

250

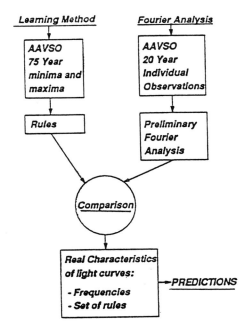

Figure 3. Complementary Approaches to Prediction

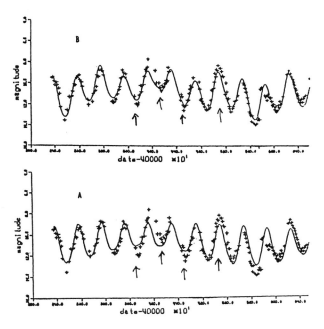

Figure 4. Fit of a Fourier development and observational data (+). Starting frequencis of fitting come only from preliminary power spectrum in A. Long-term variation suspected by learning is taken into account in B.

Each month ESOC sends to M. O. Mennessier the satellite observations of stars for which a brightness ephemeris is necessary (Barthes 1990). These observations are compared to the predicted magnitudes at Montpellier University. Every two months the AAVSO and AFOEV send to Montpellier the recent data, which are used to apply the rules to revise the predictions. Two examples of checks leading to revised ephemerides are shown in Figures 6 and 7: a phase shift for X Ceti, and a spurious frequency that was not taken out for SS Pegasi, respectively. The percentage of erroneous predictions was initially about 35%; after three revisions it is about 25%. We believe that due to the physical nature of these stars, this percentage will likely not be greatly reduced.

5. Usefulness to Stellar Physics

The characteristics of light curves (frequency and set of rules) can be used to give a better understanding of any physical problem. Let us mainly note that we can obtain information on the modes of pulsation of Miras from an examination of the frequencies (Barthes et al. 1989). The importance of this parameter in the evolutionary models is well known; it also plays an important role in regard to the respective status of semi-regulars, Miras, carbon stars, OH/IR sources, etc. But extreme care has to be taken to separate the manifestation of the stellar interior from the atmospheric phenomena (Mennessier et al. 1990).

REFERENCES

Barthes, D., Davies, P. & Mennessier, M. O. 1990, "Astrometry from Space", COSPAR Symposium, in press.
Barthes, D., Mattei, J. A., Mennessier, M. O. & Waagen, E. 1989, "From Miras to Planetary Nebulae", in Frontiéres ed., 93.
Mennessier, M. O. 1989, Applied Stochastic Models and Data Analysis 5, 253.
Mennessier, M. O. & Baglin, A. 1988, "Scientific Aspects of the Input Catalog Preparation", Turon and Torra (Eds.), 361.
Mennessier, M. O., Barthes, D. & Mattei, J. A. 1990, "Confrontation between Stellar Pulsation and Evolution", in press.
Mennessier, M. O., Burki, G. & Cordoni, J. P. 1989a, Applied Stochastic Models and Data Analysis 5, 255.
Mennessier, M. O. & Diday, E. 1989, Applied Stochastic Models and Data Analysis 5, 259.
Mennessier, M. O., Gascuel, O. & Diday, E., 1989b, Proceedings of Artificial Intelligence Techniques for Astronomy, A. Heck (ed.), Observatoire de Strasbourg, France, p. 3.
Perryman, M. A. C. et al. 1989, The HIPPARCOS Mission, European Space Agency SP-11.

Appendix: The HIPPARCOS Satellite (by Janet A. Mattei)

The High Precision Parallax Collecting Satellite (HIPPARCOS), launched by the European Space Agency in August 1989, is devoted to the precise measurement of positions, proper motions, and parallaxes of about 100,000 stars to a limiting magnitude of about 12. Through systematic scanning of the sky, 24,000 bits of data per second are sent to Earth over the satellite's three or more years lifetime, filling about 1000 high-density magnetic tapes.

The collection of this precise astrometric data is creating an enormously valuable data base which can be utilized by astronomers working in all wavelengths of the electromagnetic spectrum on studies of stellar structure and evolution, galactic motion, celestial mechanics, reference frames, and a wide variety of other astronomical applications. Mean accuracy is 0.002 arc second on each component of the position and parallax, and 0.002 arc second per year on each component of the proper motion - accuracies not achievable from ground-based observations. High precision data on position and parallaxes will allow astronomers to determine stellar luminosities, masses, and radii, and to study many aspects of stellar evolution. Proper motion observations will provide essential data for the study of the dynamics and evolution of our galaxy. For the first time direct luminosity calibrations, for which various indirect methods have so far had to be used, will be possible for regions of the Hertzsprung-Russell Diagram.

The idea for a space astrometry mission started in 1966 and evolved until 1980, when the European Space Agency adopted the mission. One hundred mostly European astronomers from 50 institutions in 12 countries are involved, forming four consortia: the Input Catalogue (INCA), two independent data analyses in France (FAST) and the Netherlands (NDAC), and Photometric Data Analysis from Star mapper, TYCO.

The HIPPARCOS observing program is guided by an Input Catalogue containing accurate photometric and astrometric information on standard stars and on objects that are being observed with the satellite. Since 1982 the INCA consortium has been measuring and compiling astrometric and photometric data on these stars and minor planets.

AAVSO Participation

Why is the AAVSO involved in an astrometric space mission? In the HIPPARCOS Observing Program 1% of the stars are variable stars, and of these about 300 of them are large-amplitude (Mira and semiregular type) variable stars of great interest for the investigation of stellar pulsation, stellar evolution, circumstellar molecular maser emission, mass loss, and chemical and dynamical evolution of the galaxy. Presently, only a limited number of long period variables have parallaxes and proper motion observations, and unfortunately, the accuracy of these values is poor.

In order to obtain precise positions of stars with the HIPPARCOS satellite, it is necessary to predict the observability windows for these stars, i.e., the times and magnitudes when these stars are brighter than the detection threshold of the satellite. To make these predictions for long period variable stars (which in general do not have strictly periodic amplitudes and phases), long-term data are needed for predicting their brightnesses and phases, and up-to-date observations are necessary for confirming and refining these predictions. The AAVSO's long and successful experience in providing important services to astronomers during satellite missions and its extensive computerized data base on large-amplitude stars led the HIPPARCOS INCA and Science Team to ask the vital support of the AAVSO for the observation of these stars with HIPPARCOS.

HIPPARCOS is obtaining precise position, parallax, and proper motion measurements of 276 long period variable stars that are in the AAVSO observing program. These high precision measurements, 100 times more accurate than any presently existing, will improve luminosity, mass, and radius determinations of these stars and will provide essential data for the study of stellar structure, evolution, and kinematics. Thus, data from HIPPARCOS will add tremendously to our knowledge of these stars.

The HIPPARCOS Input Catalogue (INCA) and Science Teams have requested the help of AAVSO observers worldwide in observing these stars during the life of the mission. AAVSO data are being used to schedule observations of these stars with the satellite.

Although this satellite had very serious initial problems with its apogee motor boosters that forced it to move in a low elliptical orbit, there is recent good news on HIPPARCOS! Due to much work by European Space Operations Center (ESOC) to optimize the operations of the satellite in its low orbit, and no major recent solar events, HIPPARCOS is running better and better. The degradation of the solar arrays is less than initially expected, and stay on or just above recent predictions. This encouraging situation leads to an extrapolated value of the lifetime of the satellite of about 1000 to 1200 days! This will make it an almost nominal mission.

HIPPARCOS has started its observations of variable stars. In the allocation of observing duration time, it is essential that it has a predicted brightness for a specific target. Thus, in making the observations of large-amplitude variable stars, it is using the brightness ephemerides prepared by INCA using the current data that you send in. I am continuously comparing the ephemerides with our recent data and alerting the HIPPARCOS Variable Star Coordinator to any discrepancies. Thus, your observations are crucial to the success of the HIPPARCOS variable star observing program. Once again, NASA, recognizing the vital role of variable star observers in this mission, is continuing its grant to the AAVSO for our participation in this project. We are grateful to NASA for this support.

DISCUSSION

Collins: Presumably the predicted light curve is a very good fit for most variables, but is there a case for quick-response e-mailed reports for those variables that do not "behave"?

Mennessier: If there is a variable that needs quick response, we alert the observers through AAVSO Alert Notices and Circulars. A few southern-hemisphere variables recently added to the program needed more observations, and southern observers were alerted to this need in this way.

INTERNATIONAL COOPERATION FOR COORDINATED STUDIES OF MIRA VARIABLES

Margarita Karovska
Harvard-Smithsonian Center for Astrophysics
60 Garden Street
Cambridge MA 02138
U.S.A.

Mira-type variables are cool giant stars (M, S, and C spectral type) on the asymptotic gaint branch of the H-R diagram which can be traced to main sequence progenitors with masses in a range from 1 to 2 $M_\odot$ ($M_\odot$ = 1 solar mass). During their evolution as Miras, these stars lose a large amount of their mass at rates of from 10^{-7} $M_\odot$/yr to a few times 10^{-6} $M_\odot$/yr (Morris 1985). Although the mechanism of mass loss is not yet completely understood, it seems that large amplitude pulsations play a crucial role.

Light curves and spectra of Mira-type stars show regular periodic variations, changing in brightness by more than 2.5 magnitudes over periods of several hundred days. Emission lines (including hydrogen Balmer lines) show amplitude and velocity variations which are consistent with pulsation (Gillet 1988). They also indicate the presence of shock waves traveling in the stellar atmosphere. Theoretical models suggest that the shock waves form in the vicinity of the photosphere as a result of the pulsation process, and then propagate outwards through the atmosphere. These shock waves extend through the upper parts of the stellar atmosphere, lifting material to large distances from the stellar photosphere. By increasing the density in regions where dust forms, the shocks create favorable conditions for substantial mass loss. Dynamical models of Mira atmospheres (Bowen 1988) indicate that dust is essential for rapid mass loss, since grains accelerated by the radiation pressure can drive large quantities of matter from the star.

The effects of pulsation on the structure of the extended atmospheres of Mira variables are also apparent in directly-measured angular diameters, which show a strong wavelength dependence (Labeyrie et al. 1977; Bonneau et al. 1982; Karovska et al. 1989). Measurements obtained in different spectral regions using occultation and speckle interferometry techniques often differ by factors of 2 to 3. Diameters measured in the absorption minima of the strong TiO molecular bands are larger than those measured in the continuum. The "true" photospheric diameter is difficult to measure due to the star's extended atmosphere, and a better understanding of its detailed structure is required.

Accurate determination of photospheric angular diameters and energy distributions for Mira variables is crucial for establishing their effective temperature scale (Ridgway et al. 1980). The effective temperature can determined from the relation

$$\text{Teff} = (4F/s\phi^2)^{1/4}$$

where F is the apparent bolometric flux received on the Earth, ϕ is the photospheric angular diameter and s the Stefan-Boltzmann constant. Obviously, an overestimate of the angular size of the photosphere will lead to a lower effective temperature.

A precisely-measured photospheric angular diameter, combined with the known distance to the star, yields its true diameter, which can then be used to determine the mode of pulsation for Mira variables (Ostlie & Cox 1986). Presently, it is uncertain whether the oscillation mode is a fundamental or a first overtone (Willson 1982; Wood 1982). Once the pulsation mode is determined, it will be possible to better define the evolutionary place of Mira variables and their relation to other types of long period variables (Willson 1982).

For the nearest Miras, an extensive two-dimensional study of the extended atmospheres can be obtained. In this sample of stars, o Ceti (the prototype of Mira-type variables) is one of the most interesting targets because of its large angular size. Speckle observations of o Ceti obtained at several epochs since 1984 show evidence for strong asymmetry in the extended atmosphere (Karovska et al. 1989; Karovska et al. 1990). Polarimetric observations have detected linear polarization from o Ceti and other Mira-type variables, indicating a departure from spherical symmetry (Boyle et al. 1986). In this program, we plan to determine the nature and the origin of the asymmetry on the o Ceti atmosphere and its possible connection to the stellar pulsation. Our aim is to monitor the asymmetry during an entire pulsation cycle in order to determine how it changes as a function of pulsation phase. We will search for similar asymmetries in other Mira-type variables in the solar neighborhood. It is crucial to determine whether asymmetries are common in Mira-type stars, since the presence of substantial asymmetries could have a major effect on the measured quantities and the model atmosphere of these stars.

The circumstellar environment of a number of Mira-type variable stars located in the solar neighborhood (within a radius of few hundred parsecs) can be studied using high angular resolution interferometry techniques (in the optical, IR and radio wavelengths). Analysis of multi-color images (400 nm to 800 nm) of the region between 2 and 5 stellar radii from the photosphere can provide important insights into the process of mass loss for these stars. Bowen (1988) suggested that in this region pulsation enhances the grain formation rate and large quantities of dust are formed and accelerated outwards. Light scattered from the grains may allow mapping of the inner boundary of the dust envelope and the detection of coarse-scale inhomogeneities (clumpiness) in the vicinity of the star. We will search for indicators of recent mass ejection events or episodes of enhanced dust formation. These observations may also aid in understanding and modeling of the observed intrinsic polarization of Mira variables. In correlation with simultaneous polarimetric observations of several Miras, we plan to obtain high angular resolution polarization observations so that we might determine the small scale nature of the scattering medium.

There are also several binary systems, including o Ceti, in which one of the components is a Mira-type star. Observations of these systems will allow a search for possible accretion disks or envelopes around these close companions (Reimers & Cassatella 1985) and a study of the wind from the cool gaint.

We propose a program of coordinated observations and modeling of the extended atmosphere and close circumstellar environment of several Miras in the solar vicinity. We expect this study to lead to better understanding of the pulsation process in Mira-type variables and to a determination of the effect of these pulsations on the atmospheric structure, temperature scale, and mass loss rates. Our sample will include M, S, and C type Miras, some of which are components of binary systems. We plan to obtain observations at different phases for up to two consecutive pulsation cycles. The photometric observations which will be obtained by A.A.V.S.O. and other observers are crucial for studying the brightness changes as a function of the pulsation phase and for predicting the brightness of studied Miras at the epochs when multi-technique observations will be carried out. High angular resolution speckle observations will be made at major telescopes located in both hemispheres. We will coordinate the speckle observations with the interferometric, spectroscopic, photometric, polarimetric, and ultraviolet observations, and the result will be combined for scientific interpretation. Numerical models generated using the observationally measured parameters will be the final product of this study.

The author wishes to thank all participants in this program of coordinated observations.

References

Bonneau, D., Foy, R., Blazit, A., & Labeyrie, A., 1982, A & A, 106, 235.
Bowen, 1989, ApJ, 329, 241.
Boyle, R.P., Aspin, C., Coyne, G.V., & McLean 1986, ApJ, 164, 310.
Gillet, D. 1988, ApJ, 190, 200.
Karovska, M., Nisenson, P., & Standley, C. 1989, BAAS, 21, No. 4 1117.
Karovska, M., 1990, these proceedings.
Labeyrie, A., Koechlin, L., Bonneau, D., Foy, R., & Blazit, A., 1977, A & A, 218, L75.
Morris, M. 1985, in Mass Loss from Red Giants, (Eds. M. Morris & B. Zukerman) D. Reidel.
Reimers, D., & Cassatella, A. 1985, ApJ, 297, 275.
Ostlie D.A., & Cox, A.N., 1986, ApJ, 311, 864.
Ridgway, S.T., & Joyce, R.R., 1980, ApJ, 235, 126.
Warner, B., 1972, MNRAS, 159, 95.

Willson, L.A. 1982, in Pulsation in Cataclysmic Variable Stars, (Eds. J.P. Cox & C.J. Hansen).
Willson, L.A. 1988, in Polarized Radiation of Circumstellar Origin, (Ed. G. Coyne).
Wood, P.R. 1982, in Pulsation in Cataclysmic Variable Stars, (Eds. J.P. Cox & C.J. Hansen).

Discussion

Le Contel. *I propose that you contact two groups, at Observatoire de la Cote d'Azur: (i) J. Guy and his group are observing, routinely, with an IR interferometer (two 1m telescopes separated by 12m on an E-W baseline; the apparent baseline varies from 4m to 12m during the night) and (ii) Cl. Mayer and his group, who make observations of lunar occultations at three wavelengths from the visible to the IR.*

Karovska.   These groups have been or will be contacted.

Boninsegna. *Another technique to measure the diameter of stars is through occultations by minor planets. These are slower than occultations by the Moon. With a high-speed photometer, the accuracy will be much better than by speckle interferometry. Presently, these events are predicted with unsufficient precision, but one day ... I invite all institutions interested in measuring the diameters of stars to keep a lookout for these phenomena.*

THE VARIABLE BINARY STAR X OPHIUCHI[1]

Dominique Proust
Observatoire de Meudon
92195 Meudon, France
Association Française des Observateurs d'Etoiles Variables

Michel Verdenet
71140 Bourbon-Lancy, France
Association Française des Observateurs d'Etoiles Variables

X Oph as a Variable Star

X Oph (M6.5e + K1 III) was discovered by Espin in 1886, and has been systematically observed since then. Values observed for maximum, minimum and period are very discordant in the literature, e.g. 6.8 to 8.8 in 335.4 days (Townley et al. 1928), 5.9 to 9.2 in 335.1 days (1st edition of the GCVS) and < 5.9 > to < 9.2 > in 328.85 days (most recent edition of the GCVS.)

This star has the advantage of being quite bright, so that the light curve leads to determinations of maximum with good accuracy. However, the main problem with X Oph is that its real minimum brightness is in fact unknown since the star becomes fainter than its companion, as was shown spectroscopically by Keenan et al. (1974). Fig. 1 shows its variation between 1980 and 1989. X Oph is the only Mira-star among a sample of forty to present a single frequency in the power spectrum of its light curve with $\nu = 0.00301$, $\alpha = 0.68$ $\sigma_{o-c} = 0.28$ between 1979 and 1984, and $\sigma_{o-c} = 0.26$ in 1985-86 (Mennessier et al. 1989): this frequency can be explained by the constant light minimum from the companion and can be considered as a criterion of binarity.

We have computed from the literature the observed periods during 113 cycles, from 1886 until 1989. Fig. 2 shows the plot of individual period and maximum brightness as a function of the cycle; some maxima are missing, corresponding to a solar conjunction. Note however the increasing scatter since 1950 (cycle 70), and the general trend of decreasing maximum brightness. The average period using all the cycles is P = 333.47 days. The histogram in Fig. 3 shows an asymmetry of the period distribution with an excess between 335 and 350 days. The Gaussian fit gives P = 334.15 days with $\sigma = 12.3$ and $X^2 = 15.36$: such results cannot support the reality of a systematic increasing period; within the next century one should have a valuable diagnostic! If we consider the O-C diagram, no more conclusions can be reached since a continuous function represented by a Fourier fitting reflects only the cumulative error.

---

[1] based on observations made at Haute-Provence Observatory, CNRS

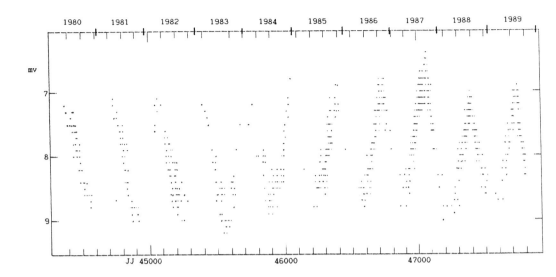

Fig. 1: Light curve of X Oph 1980 - 1989.

The spectrum of X Oph has been extensively studied by several authors; as usual for this kind of star, it is incredibly complex, dominated in the blue regions by atomic absorption lines, in the visual and near infrared by bands of metallic oxide molecules, and in the infrared beyond 1.5μ by innumerable lines from the rotation-vibration transitions of CO and H_2O. In addition to these absorption features, emission lines of various descriptions are present: hydrogen lines of the Balmer, Paschen and Brackett series are strong in emission during more than half of the cycle, from just before maximum to approximately the time of minimum light (Wing 1979).

Among the spectral type determinations, Keenan et al. (1974) have established a table of mean spectral types at normal maximum brightness based on the intensities of the CaI $\lambda 4226$, Cr I $\lambda 4254$, SrII $\lambda 4077$, FeI $\lambda 4063$-4071 absorption lines, H_δ and H_β in emission, and the strongest blue bands of AlO and ZrO. From a series of spectra taken between Aug. 1966, and Mar. 1968, they conclude for X Oph a mean spectral type at normal maximum being M6.5e corresponding to an effective temperature close to 2550° and a radius $\log(R/R_\odot) \approx 2.5$.

Despite the difficulties and uncertainties that plague the models for Miras, there are some quantities which appear to be reasonably model-independent. The

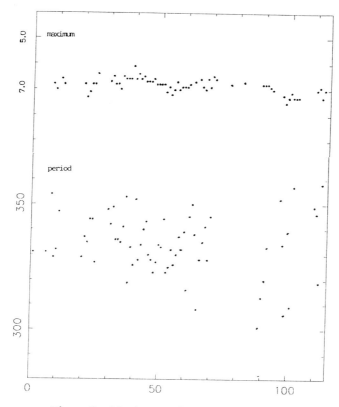

Figure 2: Maxima and period of X Oph

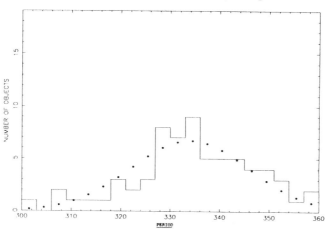

Figure 3: Period distribution of X Oph

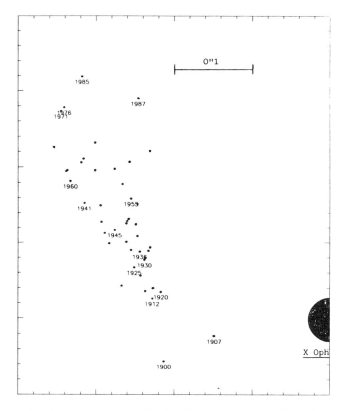

Figure 4: Apparent motion in the binary system X Oph since 1900.

fundamental mode period - mass - radius relation is (Willson 1981):

$$\log P = -1.96 - 0.67 \log (M/M_\odot) + 1.87 \log (R/R_\odot)$$

leading to a mass $M = 1.9 M_\odot$. Such a value corresponds with results published by several authors. A new approach to the mass determination can be made with the binary properties of X Oph.

The Binary System X Oph

X Oph is one of the three well-known long-period variables (with Mira and R Aqr) to belong to binary systems, and none of the three can be considered entirely suited to a determination of mass.

The companion of X Oph is a K1 III giant (Proust et al. 1981) for which more than 150 individual observations were made since 1900. These data serve to define no more than a 65° arc of the apparent orbit, but this arc is known with relatively high precision. The data were collected in Fernie (1959), Baize (1980) and Couteau (1990). We added unpublished measurements made in 1971 by one of us (D.P.) with the 76 cm refractor at Nice Observatory (Fig. 4). The orbital elements are as follows (Baize 1978):

$$P=485.3 \text{ years}, T=1873.6, a=0.34'', e=0.715, i=150.6°, \Omega=72.9°, \omega=139.9°$$

The recent measurements are in agreement with the above orbit within only 3°. Note that McAlister (1978) predicted a probable third component associated with A or B at $\theta = 42.6° \pm 180°$ and $\rho = 0.053''$.

These results lead us to determine the stellar masses, using Kepler's third law:

$$a^3/P^2 = m_1 + m_2$$

where a is expressed in astronomical units; distance is calculated using the relation between period and absolute magnitude at the mean maximum as given by Foy et al. (1975). The observed visual magnitudes were corrected for interstellar absorption. We obtain a distance of 300pc for X Oph, corresponding to a sum of the masses of $4.5 M_\odot$. Such a result is in a very good agreement with masses predicted from stellar evolution.

The absolute magnitude of a typical K1 giant is $M_v = +1.0$. The classical formula:

$$M = m - 5 \log (D/10)$$

with D expressed in pc, leads to an apparent magnitude $m_v = 8.4$. Considering the interstellar absorption used above of 2.3 magnitude/kpc, we can derive a visual magnitude $m_v = 9.1$: at minimum, the light contribution of X Oph becomes negligible. This is supported by the colour index of the system at minimum light: $B - V = 1.29$. Taking the intrinsic colour as being 1.08 in the Johnson system, the excess is $E_{B-V} = 0.19$, whence $A_v = 3E_{B-V} = 0.63$, which agrees very well with the above determination.

Such results place the star on the giant branch of the H-R diagram of classical clusters such as NGC188 or M67. If we consider theoretical fluxes, and absolute magnitude, we obtain from evolutionary tracks (Vandenberg 1983) a mass $M = 3 \pm 1 M_\odot$. It follows that X Oph should have a mass $M = 1.5 \pm 1 M_\odot$, corresponding to a mass ratio $M_M/M_K = 0.5$, which is consistent with the value of 0.75 (Fernie 1959) obtained from radial velocity measurements. Such a result is also in a very good agreement with the above spectroscopic determination leading to $M = 1.9 M_\odot$.

An estimate of the mass loss rate can be made for X Oph using the relation of Reimers (1975, 1977):

$$dM/dt = -A \cdot L \cdot R/M$$

where $A = 4.10^{-13}$ $M_\odot$/year. Using the above data with a temperature $T = 2550°K$ and an average mass $M = 1.7 M_\odot$ we obtain:

$$\log(-dM/dt) = 5.9$$

which is very consistent with the value of 6.05 found by De Jager et al. (1988): the mass loss amounts to 1.26×10^{-6} solar mass per year. Such a result is within the determinations of Willson (1981) ranging from 10^{-5} to 2×10^{-6} $M_\odot$/year.

X Oph can be considered as one of the Mira prototypes. Its binary nature has in no way affected the nature of the long-period variable since its physical characteristics, mass loss rate and possible period change are representative of the evolution of a $1 < M < 3 M_\odot$ star along the asymptotic giant branch of the H-R diagram. Such stars develop into Mira stars towards the end of their evolutionary tracks to become white dwarfs ($M = 0.6$ $M_\odot$) either directly, or through the intermediate stage of planetary nebulae.

References

Baize, P. 1980, A&AS, 39, 83.
Couteau, P. 1990, private communication.
De Jager, C., Nieuwenhuijzen, H. & Van der Hucht, K.A. 1988, A&AS, 72, 259.
Fernie, J.D. 1959, ApJ, 130, 611.
Foy, R., Heck, A. & Mennessier, M.O. 1975, A&A, 43, 175.
Keenan, P.C., Garrison, R.F., Deutsch, A.J. 1974, ApJS, 28, 271.
McAlister, H.A. 1978, ApJ, 225, 932.
Mennessier, M.O., Burki, G. & Cordoni, J.P. 1989, Applied Stochastic Models and Data Analysis, 5, 255.
Proust, D., Ochsenbein, F. & Pettersen, B.R. 1981, A&AS, 44, 179.
Reimers, D. 1975, Commun. 19. Colloq. Int. Astrophys. Liége; Mèm. Soc. R. Sci. Liége, Ser 6, 8, p369.
Reimers, D. 1977, A&A, 61, 217.
Townley, S., Cannon, A.J. & Campbell, L. 1928, Annals Harv. Coll. Obs., 79.
Vandenberg, D. 1983, ApJ, 51, 29.
Willson, L.A. 1981, in Pulsation in Classical and Cataclysmic Variable Stars, eds J.P. Cox and C.J. Hansen, p. 269.
Wing, R.F. 1979, Contrib. Perkins Obs., series II, #80.

CHAPTER 7

CATACLYSMIC VARIABLE STARS

SYMBIOTIC VARIABLE STARS

Joanna Mikolajewska
Institute of Astronomy
Nicolaus Copernicus University
PL-87100 Torun, Poland

1. Some History

The existence of symbiotic stars was first noted by Williamina P. Fleming who included R Aqr and RW Hya in her list of "stars with peculiar spectra", and Annie Cannon who isolated a group of red stars with bright H I and He II emission lines (Z And, CI Cyg, SY Mus, AG Peg) during her work on the HD Catalog. These discoveries, however, went unrecognized till Merrill and Humason (1932) rediscovered CI Cyg, RW Hya and AX Per as peculiar M stars with strong He II 4686 emission lines. The simultaneous presence in a single stellar object of low-temperature absorption features (TiO bands, neutral metals) and emission lines that require high excitation conditions in the 1930s was unique, and the newly discovered *stars with combination spectra* have become attractive targets for many astronomers. Soon after, examination of the photometric archives at Harvard College Observatory revealed that some of these objects (e.g. Z And, CI Cyg, AX Per) show more or less regular light changes with periods of 600-900 days as well as occasional 2-3 mag nova-like eruptions. Simultaneously, extensive spectroscopic studies by Merrill at Mt. Wilson Observatory, Swings and Struve at McDonald and Yerkes Observatories, and many other astronomers showed that the outbursts are accompanied by dramatic spectral changes: a strong blue continuum overwhelms the late-type absorption features, most of the high-excitation lines vanish as well, while the bright H I and He I lines usually develop P Cyg absorption components.

The observational characteristics and the outburst behaviour pointed at binary nature of most if not all symbiotic stars but verification of the idea by observations of eclipses and determinations of spectroscopic orbits has come only recently. Nevertheless, in 1941, P. Merrill proposed to call Z And, CI Cyg, AX Per and related systems symbiotic stars, clearly referring to their binary appearance.

By the late 1960s, astronomers had collected an enormous amount of optical data describing the behaviour of the symbiotic stars; their underlying physical nature, however, remained uncertain. In the 1980s, the symbiotic stars became fashionable again. Advances in technology have made possible observations spanning the electromagnetic spectrum from low-frequency radio to hard X-rays, and resulted in essential improvement in understanding the very complex symbiotic phenomena observed in the optical. The present paper presents a brief summary of our current knowledge of the symbiotic stars. Mostly due to the limited space

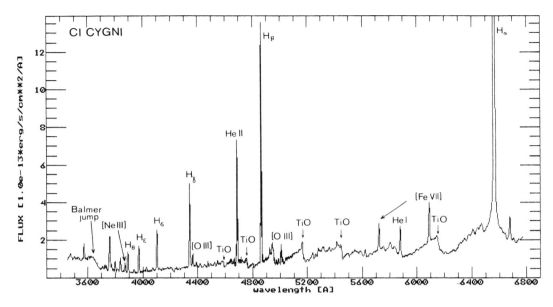

Figure 1: *Optical spectrum of CI Cygni*

available here I have given references only to specific cases and very recent discoveries, and I apologize to all whose works I have neglected. For the most comprehensive reviews the interested reader is referred to the proceedings of the 1981 and 1987 international meetings on the subject (Stencel 1981; Friedjung & Viotti 1982; Mikolajewska et al. 1988) and the only book published thus far on these stars (Kenyon 1986) where also a complete bibliography is given.

2. Multifrequency Observations

A. Optical data

Because it is in the optical that the symbiotic stars were first recognized, and still are classified, it is appropriate to start with an optical spectrum of a typical object, to show the defining features of the class. Such a spectrum is presented in Fig. 1, and the basic characteristics are:

1. absorption features (TiO, VO, Ca I and Na I), and an associated red continuum observed in red giants;

2. the Balmer jump in emission and strong emission lines from relatively highly ionized species (H I, He I, He II, [O III], etc.) found in planetary nebulae, and a blue continuum.

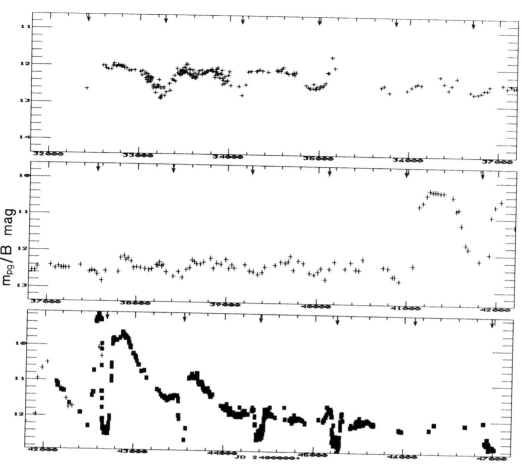

Figure 2: m_{pg}/B light curve of CI Cyg. Arrows indicate times of eclipses from the ephemeris (Kenyon et al. 1990): JD Min = 2442685.5 + 855.25 E.

The spectrum of a symbiotic star in outburst is usually similar to that of an A-F supergiant with additional emission lines of H I, He I and singly ionized metals.

Periodic photometric variations, with periods of 200-1000 days and amplitudes tending to increase towards short wavelengths, were observed in many symbiotics. Although such behavior was not observed in other long-period red variables, it was generally not associated with binary motion until the late 1960s, when Hoffleit (1968) noted the mean light curve of CI Cyg resembled that of an eclipsing binary, and Belyakina (1970) successfully interpreted the light curve of AG Peg in terms of the reflection effect. The eclipsing nature of CI Cyg was verified in a series of UBV observations obtained by Belyakina (1989 and references therein). The m_{pg}/B light curve of CI Cyg is presented in Fig. 2. Other eclipsing symbiotic stars are: BF Cyg (Mikolajewska, Kenyon & Mikolajewski, 1989), CH Cyg (Mikolajewski, Tomov & Mikolajewska 1987), SY Mus (Kenyon et al. 1985), AR Pav

(Mayall 1937), and AX Per (Mikolajewska 1988). By now, 21 systems have known binary periods.

The very low radial velocity amplitudes expected in a binary containing a red giant, $K_g \sim$ 5-10 km/s, makes direct identification of the symbiotic stars as binary systems a very difficult task. Nevertheless, modern observations with photon counting detectors and cross-correlation analysis have demonstrated that all bright symbiotics are low-mass ($M_{total} \sim$ 2-3 $M_\odot$) double stars (Garcia & Kenyon 1988).

B. Infrared Data

Symbiotic stars divide into two major classes according to their near-infrared properties: the S (stellar) type and the D (dusty) type. The majority of systems (~80%) belong to the S class and have near-IR colors consistent with stellar photospheric temperatures of 3000 K to 3500 K. A few other objects are the D-type systems and their near-IR colors indicate the presence of thermal radiation from hot (T ~ 1000 K) dust. IR photometric monitoring has demonstrated that these D-type objects show periodic pulsation ($\Delta K \sim 1^m$, P ~ 300-600 days) resembling Mira variables (e.g. Whitelock 1988). Allen (1983) and Kenyon, Fernandez-Castro & Stencel (1986) showed that the IR continua of several D-type symbiotics are consistent with heavily reddened Mira variables (extinctions at 2μm, $A_K \sim 0.5\text{-}2^m$).

More than 50% of known symbiotics were detected by the Infrared Astronomical Satellite. The IRAS data at 12-60 μm were used for studying mass loss rates from the cool components of symbiotic stars (e.g. Whitelock 1988, Kenyon, Fernandez-Castro & Stencel 1988). This data imply mass loss rates, $M \sim 10^{-5}$ $M_\odot$/yr for the D-type systems, which is somewhat larger than that of most single Miras. It also appears from the 12 μm excess observed in most S and D-type symbiotics (Kenyon 1988) that they generally display more dust emission and lose mass more rapidly than do single red giant stars.

The division of symbiotics into S and D types is likely to hold a key to an overall understanding of the symbiotic phenomenon. Many observational characteristics (e.g. radio emission, outburst behaviour), and physical parameters (e.g. the nature of the cool component, size and density of the nebular region, orbital periods: ~ 500-1000 days, and > 15 years, for S and D systems, respectively) are correlated with the membership in the particular population. At present, the distinction seems to be one of orbital separation: whether the cool star has been permitted to evolve to a Mira state with substantial wind and dust production.

C. Ultraviolet Spectra

The launch of the International Ultraviolet Explorer revolutionized the study of symbiotic stars, and provided direct UV spectra with very intense blue continua

(Fig. 3a) that can be understood simply in terms of the Rayleigh-Jeans tail of a hot blackbody and $fb + ff$ emission from an ionized nebula. However, a few symbiotics, for instance CI Cyg (Fig. 3b), have relatively flat continua more typical of A or B-type stars. It is obvious that an A or B star is not able to power the observed emission spectrum. This apparent inconsistency of the UV continuum and emission line spectrum can be overcome if such continua arise from an accretion disk surrounding a compact star. Kenyon & Webbink (1984) demonstrated that the flat UV continua can be produced in an accretion disk around a low mass main sequence star. The required accretion rate, $M_{acc} \sim$ a few $\times 10^{-5} M_\odot$/yr, is fairly high, and the cool giant should fill or nearly fill its tidal lobe. It is interesting that main sequence accretors provide the best-fit solution for four systems (CI Cyg, YY Her, AR Pav and AX Per). They all belong to the S-class and three of them are eclipsing binaries with orbital periods: $\sim 605^d$ (AR Pav), $\sim 682^d$ (AX Per) and $\sim 855^d$ (CI Cyg), respectively.

Emission lines from highly ionized species such as He II, C III, C IV, Si III, Si IV and N V, and in a few cases O I and Fe II, are fairly strong in UV spectra of symbiotics, confirming the presence of large ionized nebulae inferred from optical data. These features can be broad - full width $\sim$ 500-1000 km/s (e.g. AG Peg, BF Cyg, CH Cyg), or narrow, $\sim$ 100 km/s (e.g. CI Cyg, AX Per). The emission lines may also be used as a probe of the physical conditions in the symbiotic nebulae. Typical results for the electron temperatures, $T_e \sim$ 10000-20000 K, show that ionization is done by radiation and not by collisions. The density diagnostics indicate that the gas is much denser, $n_e \sim 10^6$-10^{10} cm^{-3}, than in typical planetary nebulae or H II regions, $n_e < 10^4$ cm^{-3}. The nebulae of the D-type symbiotics tend to have lower densities.

D. Radio Emission

If the ionized nebulae surrounding the symbiotic binary system is large enough, we might expect measurable radio emission. By now radio thermal bremsstrahlung has been detected at centimeter wavelengths from about 35 symbiotic stars (Seaquist 1988). The radio emission is associated with the amount of dust in the stellar envelope: the D-type symbiotics exhibit the largest fluxes in both the radio and the IR range. The radio luminosities are also correlated with the red giant spectral types, consistent with the mentioned correlation with IR dust emission, since Miras present in the D-type systems have the latest spectral types. It is apparent that the D-type symbiotics have very extensive ionized regions, with $R_{6cm} \sim$ 100-1000 AU.

Radio maser emission (SiO and OH) is characteristic of many Mira-like variables. Two Mira symbiotics, R Aqr and H1-36, have been detected as SiO sources, but the hot components of other symbiotics seem to maintain conditions which are unfavorable for such maser emission (Allen et al. 1989).

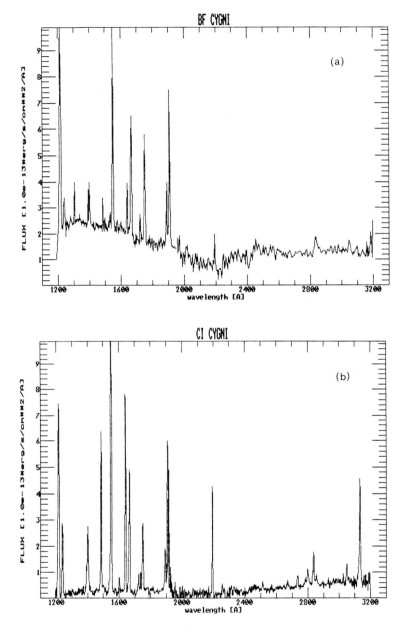

Figure 3: Ultraviolet spectra for BF Cygni (a) and CI Cygni (b).

Some of the more intense sources (~10) show complex asymmetric structure at arcsec or subarcsec resolution. Taylor (1988) divided them into two categories: ejecta and stellar winds. Among the ejecta-type there is a strong tendency for bipolar or jet morphology. It is interesting that the objects with bipolar or jet morphology often show broad structure, ≥ 500 km/s, in the UV emission line profiles (e.g. CH Cyg, AG Peg, HM Sge) indicating that these systems may also possess hot star winds.

E. X-rays

The most spectacular X-ray symbiotic star is undoubtedly GX1+4/V2116 Oph, which is a neutron star accreting from the wind of an M giant companion. Throughout the 1970s, GX1+4 was one of the brightest hard X-ray sources, and its period derivative was among the largest known for a neutron star (e.g. McClintock & Leventhal 1989 and references therein). McClintock and Leventhal have also suggested that GX1+4 is the intense 511 keV positron annihilation-line source detected from the direction of the Galactic center.

Four symbiotic stars have been detected as soft X-ray sources with the Einstein satellite (Allen 1981; Anderson, Cassinelli & Sanders 1981). Three of them, HM Sge, V 1016 Cyg and RR Tel have shown slow-nova eruptions in the past 40 years, while AG Dra was observed during the rise to the 1981-1985 series of eruptions. The possible interpretation for the X-rays involve thermonuclear events on wind-accreting white dwarfs (e.g. Allen 1981) and/or shocks in stellar winds (e.g. Kwok & Leahy 1984). Two additional detections were made with EXOSAT: R Aqr and CH Cyg, where the high energy photons are probably associated with the jets (Viotti et al. 1987, Leahy & Taylor 1987).

Most of the symbiotics (~70%), however, observed with Einstein and EXOSAT have not been detected as X-rays sources (e.g. Kenyon 1986). Local absorption of the X-rays by the nebula probably prevents many symbiotic stars from being strong X-ray emitters.

3. Physical causes that give rise to the symbiotic phenomenon

The multifrequency observations reviewed in the preceding section require symbiotic stars to be <u>long period interacting binary systems</u>. We must, however, emphasize that there are several types of interaction that can stimulate the apparition classifiable as symbiotic object. Allen (1988) proposed the following subclasses of symbiotic stars:

1. A main sequence star accretes via Roche-lobe overflow from an M giant. The UV radiation is produced in an accretion disk. Changes in the mass transfer rate and/or disk instabilities may cause outbursts observed in the optical. Kenyon & Webbink (1984) demonstrated that for M_{acc} ~ a few x 10^{-4} $M_\odot$/yr the continuum produced in the disk resembles that of an A-F supergiant. The best studied example

is CI Cyg (Kenyon et al. 1990).

2.   A white dwarf accretes from the M giant (or Mira) wind. The UV luminosity is provided by steady (or nearly steady) nuclear burning of the accreted material. Small variations in M_{acc} can transform the hot UV source into a star resembing an A-F supergiant and give rise to an optical outburst. A probable example is BF Cyg (Mikolajewska, Kenyon & Mikolajewski 1989).

3.   A white dwarf that has accreted from the wind of an M giant or Mira and so accumulated unburnt hydrogen, undergoes a shell flash. As a result we have an ultra-slow novae, like AG Peg, HM Sge or V1016 Cyg (Kenyon 1986, Mikolajewska et al. 1988).

4.   A neutron star accretes from the M giant wind. The only known case is the aforementioned GX1+4/V2116 Oph.

The interaction between the components of a symbiotic system can be further complicated by, for instance, the presence of a magnetic white dwarf as in the case of CH Cyg (Mikolajewski et al. 1990).

It is also possible that a planetary nebula may be forming alongside a red giant. Simple statistical arguments (Mikolajewski, Kenyon & Mikolajewski 1989) suggest that only a very small portion of symbiotic stars can be such systems, however.

4. Why are astrophysicists interested in symbiotic stars?

There are many reasons. First, as in all binary systems, they provide direct information about the fundamental parameters of the stars involved, such as masses and radii, that are essential to understanding stellar evolution. Second, they interact. Interactions between the stars in a binary system are of great interest in astrophysics today, and symbiotics give us an excellent opportunity to study these processes under the very extreme physical conditions found in these systems. Thus, symbiotics tell us about such basic physical processes as: mass loss from red giants, accretion onto compact stars (main sequence stars, white dwarfs, neutron stars) and the evolution of nova-like eruptions in very wide binary systems, and radiative processes in gaseous nebulae. Third, the symbiotics are among the binaries with the longest known periods. Thus, each component has enough time to fulfill its evolutionary destiny before their interaction begins and a symbiotic star is born. The variety of possible combinations giving rise to the symbiotic phenomenon, combined with knowledge of masses and radii of the stars involved, make symbiotics very important for understanding the last stages of stellar evolution. Finally, last but not least, symbiotic stars simply do exist and it is natural that we try to learn as much as possible about them.

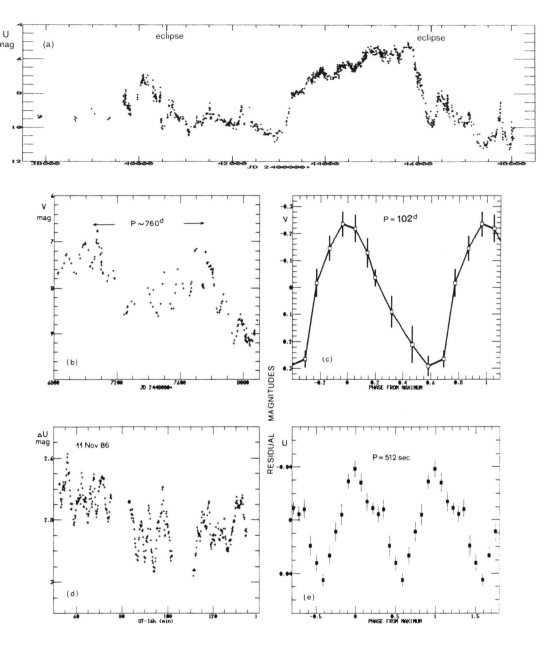

Figure 4: Variable phenomena in CH Cygni: (a) U light curve from the recent series of outbursts. Arrows mark times of eclipses of the hot component. (b) V light curve of the M giant. (c) Residuals of the data shown in panel (b) binned and folded with the 102^d pulsation period of the giant. (d) Rapid variability in U light. (e) Residual normal points of the data plotted in panel (d) folded with the 512s period (Mikolajewski et al. 1990).

5. How can amateur astronomers help?

Amateur astronomers have played a very important role in photometric monitoring of variable stars throughout this century, and symbiotics are especially grateful targets for such research. Symbiotic stars are unusual among variable stars for they fluctuate in several different ways. The record for this type of behavior may he held by CH Cyg, one of the best studied objects (Fig. 4).

Patient monitoring of light curves of symbiotic stars can reveal such variable phenomena as:

- binary motion - manifested by eclipses of the hot component by the giant (e.g. CI Cyg - Fig. 2) or by the reflection effect (e.g. AG Peg); expected periods: $P < 1000$ days (S-type) and $P \sim 10$ to 100 years (D-type);

- active states or outbursts at irregular intervals of time, lasting for a few years up to several decades;

- pulsations (or semi-regular variations) of the cool giant with typical periods, $P \sim 40$ to 600 days;

- rapid variability, flickering, with typical timescales, $\Delta t \sim$ minutes;

- solar-type cycles of the late-type giants (possible example AR Pav);

- obscuration events by dust formed in the late-type component (or system?) envelope (e.g. R Aqr);

At present, about 140 symbiotic binaries are known (Kenyon 1986). Only 21 systems ($\sim 15\%$) have known orbital periods, and we desperately need more. Although it is now commonly accepted that the cool components in the D-type symbiotics are Mira variables, only $\sim 50\%$ of them have known pulsation periods. Given the very complex pattern of variability of most symbiotics, long-term photometric monitoring by amateur astronomers may help significantly.

I would like to thank P.P. Eggleton from the Institute of Astronomy at University of Cambridge, where a portion of this work was completed, for many stimulating discussions and his gracious hospitality. I am also very grateful to M. Milokajewski for preparing Figure 4. Finally, I thank the AAVSO for their support.

References

Allen, D.A., 1981, MNRAS, 197, 739.
Allen, D.A., 1983, MNRAS, 204, 113
Allen, D.A., 1988, in "The Symbiotic Phenomenon", eds. Mikolajewski et al., Kluwer, Dordrecht, p.3.
Allen, D.A., Hall, P.J., Norris, R.P., Troup, E.R., Wark, R.M., Wright, A.E., 1989, MNRAS, 236, 363.
Anderson, C.M., Cassinelli, J.P., Sanders, W.T., 1981, ApJ, 247, L127.
Belyakina, T.S., 1970, Astrofizika, 6, 49.
Belyakina, T.S., 1989, Izv. Krym. Astrofiz. Obs., 83 in press.
Friedjung, M., Viotti, R. (eds), 1982, "The Nature of Symbiotic Stars", Reidel, Dordrecht.
Garcia, M.R., Kenyon, S.J., 1988, in "The Symbiotic Phenomenon", eds. Mikolajewski et al. Kluwer, Dordrecht, p.27.
Hoffleit, D., 1968, Irish Astr. Bull., 8, 149.
Kenyon, S.J., 1986, "Symbiotic Stars", Cambridge Univ. Press.
Kenyon, S.J., 1988, AJ, 96, 337.
Kenyon, S.J., Fernandez-Castro, T., Stencel, R.E., 1988, AJ, 95, 1817.
Kenyon, S.J., Webbink, R.F., 1984, ApJ, 279, 252.
Kenyon, S.J., Michalitsianos, A.G., Lutz, J.H., Kafatos, M., 1985, PASP, 97, 268.
Kenyon, S.J., Oliversen, N.A., Mikolajewska, J., Mikolajewski, M., Stencel, R.E., Garcia, M.R., Anderson, C.M., 1990, AJ, submitted.
Kwok, S., Leahy, D.A., 1984, ApJ, 283, 675.
Leahy, D.A., Taylor, A.R., 1987, A&A, 176, 262.
Mayall, M.W., 1937, Ann. Harv. Coll. Obs., 105, 491.
McClintock, J.E., Leventhal, M., 1989, ApJ, 346, 143.

Merrill, P.W., Humason, M.L., 1932, PASP, 44, 56.
Mikolajewska, J., 1988, in "The Symbiotic Phenomenon", eds. Mikolajewska et al. Kluwer, Dordrecht, p.303.
Mikolajewska, J., Kenyon, S.J., Mikolajewski, M., 1989, AJ, 98, 1427.
Mikolajewski, M., Tmov. T., Mikolajewska, J., 1987, ApSS, 131, 733.
Mikolajewski, M., Mikolajewska, J., Khudyakova, T.N., 1990, A&A, 235, 219.
Mikolajewski, M., Mikolajewska, J., Tomov, T., Kulesza, B., Szczerba, R., Wikierski, B., 1990, Acta Astr., 40, 129.
Seaquist, E.R., 1988, in "The Symbiotic Phenomenon", eds. Mikolajewska et al., Kluwer, Dordrecht, p.69.
Stencel, R.E., (ed), 1981, "Proc. N. Amer. Workshop on Symbiotic Stars", JILA NBS and Univ. of Colorado.
Taylor, A.R., 1988, in "The Symbiotic Phenomenon", eds. Mikolajewska et al., Kluwer, Dordrecht, p.77.
Viotti, R., Piro, L., Friedjung, M., Cassatella, A., 1987, ApJ, 319, L7.
Whitelock, P.A., 1988, in "The Symbiotic Phenomenon", eds. Mikolajewska et al., Kluwer, Dordrecht, p.47.

Discussion

Williams: You mentioned that there were a few symbiotic stars that are known to have accretion disks, and also that four or five symbiotic stars are thought to have bi-polar structures (possible jets?). Are these stars in the same class as the symbiotic stars with accretion disks?

Mikolajewska: <u>All</u> symbiotic nebulae resolved at radio wavelengths show bipolar morphology. Such axial symmetry might reflect the presence of accretion disks. On the other hand, none of the symbiotics containing a <u>disk accreting main sequence stars</u> (case 1, section 3) has not been resolved at radio wavelengths; they are relatively weak radio sources. However, accretion disks can be also formed from wind accretion.

Mahmoud: What are the limits of the optical region used to define the symbiotic stars?

Mikolajewska: Symbiotic stars were first isolated based on their optical spectra in the wavelength range between 3400 and 5100 A (HD survey). Now, we can use optical spectra in the range $\lambda\lambda$ 3400 - 7000 A.

THE BENEFIT OF AMATEUR OBSERVATIONS FOR RESEARCH IN DWARF NOVAE

Constanze la Dous
NSA/GSFC/NSSDC Code 633
Greenbelt, Maryland 20771 USA

1. Introduction

Dwarf novae are members of the class of cataclysmic variables; this class also includes novae, recurrent novae, and nova-like stars. The defining characteristic of all these objects, with the exception of nova-like objects, is that at times they undergo sudden brightness increases by several magnitudes. It is believed that all cataclysmic variables are close interacting binary stars, and that the same model describes their basic nature. The cause of the outburst, however, is a thermonuclear explosion in the case of novae and recurrent novae, while it is some event connected with the accretion disk in the case of dwarf novae; nova-like objects are assumed to be a special case of dwarf novae.

1.1 The Roche Model

It is assumed that all cataclysmic variables are binary systems in which the secondary, i.e. the less massive component, is a cool main sequence star (spectral type K or M), while the primary star is a white dwarf. Furthermore, the stars are so close to each other that the secondary star just fills its Roche lobe, that is the critical volume within which material clearly belongs to this star. As in the course of time it expands slightly, for instance due to evolutionary processes, its surface becomes a little larger than this critical volume, and this in turn means that the material gets lost to the star. Most easily this happens in the neighbourhood of the white dwarf (since at the point where the two critical volumes meet there is no gravitational attraction holding the material back to the secondary star), and thus a constant, normally small stream of material is spilled into the volume of the white dwarf. It is one of the characteristic properties of a white dwarf to be very small, though rather massive, and consequently this star does not nearly fill its Roche volume but is just a little dot of material in its center. Material entering its Roche lLobe is attracted to the center of the white dwarf, but, due to conservation of angular momentum, it cannot fall onto the star but rather forms an accretion disk, and in this disk it eventually spirals ever closer to the surface of the white dwarf. As the material approaches the center it is heated up, an effect which is crucial for the occurrence of dwarf nova outbursts.

1.2 Amateur Contributions

Now I want to turn to dwarf novae more specifically and I want to present them from the point of view of contributions of amateur observations to their understanding. These contributions have been in the areas of object identification and classification, of providing long-term outburst light curves which provide important clues to the structure, stability, and dynamics of the accretion disk, as well as to the outburst mechanism; then they have been extremely helpful in scheduling satellite observations. Finally, it often has been possible to after the observations determine the precise position of the obtained data in the outburst light curve and/or obtain the outburst history for some time before and after the observations, which then often gives important clues as to the physical processes involved.

I will concentrate on three main aspects: on the analysis of outburst light curves, in the UV delay, and, finally I want to present one sort of observation that has not generally been carried out by amateurs so far, but which I think might be just at the limit of possibilities, and, if done, would be of immense help for the further understanding of these objects: this is orbital photometry of dwarf novae. I will not be able to go into too much detail of every individual aspect, but for more extensive coverage see la Dous, 1990 ('Cataclysmic Variables and Related Objects, NSA/CNRS Monograph on Non-Thermal Phenomena in Stellar Atmospheres, M. Hack (ed.), p. 15 - 256, in press; copies will be sent to the heads of variable star observing groups of the countries from which members attended this meeting).

2. The Outburst Light Curves

Dwarf novae are defined by the optical appearance of their outburst light curves: At semi-periodic intervals of time they all undergo outbursts of some 3 to 5 mag amplitude; the "periods" range between some 10 and 100 days; each "period" is characteristic for each object. Commonly three types of dwarf novae are distinguished, again according to features seen in the outburst light curve. One type is the *SU UMa stars*; these, in addition to having 'normal' outbursts, also undergo, in general much less frequent, superoutbursts. These are somewhat brighter than the normal outbursts and they last about ten times as long. A number of other phenomena, which do not show up in the outburst light curve, are connected with superoutbursts. The second class are the *Z Cam stars*; these, in addition to having rather frequent "normal" outbursts, occasionally cease all apparent activity at an intermediate brightness level, normally, though not always, on decline from a normal maximum. These standstills can last for just a few days, or up to years, after which time the system returns to the quiescent level and resumes normal activity. The occurrence of the next standstill is unpredictable, it can occur as soon as after the next maximum. And finally, there is the class of *U Gem stars*; these are objects that are not known to be either SU UMa or Z Cam stars. The membership of a system in the class of the SU UMa or the Z Cam stars seems to be mutually exclusive, i.e. no object is known which at the same time possesses the characteristic qualities of both groups. However, for some systems the classification as a U Gem star is a reflection of incomplete observational records; a number of stars that originally were classified as U Gem objects, now are known to be SU UMa or Z Cam stars. But certainly there are objects that are genuine U Gem stars in that they clearly do show neither superoutbursts nor standstills; the best known one is SS Cyg which has the most complete observational record of all dwarf novae.

The classification given above is the one currently in most use, but in the past different names as well as different classification systems were used (noticeably the one introduced by M. Petit which distinguished seven sub-classes); and also these days what above I called U Gem stars sometimes are referred to as SS Cyg stars, or both names are used as generic names meaning the entire class of dwarf novae. The current classification of individual objects largely relies on the efforts of H. Ritter (1987 A&AS 70, 335; new edition A&AS, in press).

As for the shapes of outbursts, in most, if not in all, objects so-called "normal" outbursts with rise times on the order of one day can be distinguished from "anomalous" outbursts where the rise lasts for several days, so that rise and decline are of about equal duration. In addition, short-lasting and long-lasting outbursts occur, but the duration of an outburst does not appear to be related to the rise time.

2.1 Outburst Theories

There are two competing models that try to explain why outbursts occur in dwarf novae: the *transfer instability model*, and the *disk instability model*. In the transfer instability model it is assumed that at all times there is a small stream of material coming from the secondary star into the Roche volume of the primary, but at semi-periodic intervals of time, the atmosphere of this star becomes a little unstable and for a short time much more material is transferred. The accretion disk cannot cope with this excess material adequately, but collapses; in the course of this collapse, suddenly a lot of material is transported through the disk towards the white dwarf; it heats up, brightens, and an outburst occurs. When all excess material is on its way to the white dwarf, the peak of the outburst is over, the system cools down, and the whole process starts again.

In the case of the disk instability model, the concept is that, at all times, the stream of gas into the Roche lobe of the white dwarf is more or less constant, but during the quiescent phase of the outburst cycle material is not properly transported onto the white dwarf but rather stored in the outer part of the accretion disk, far away from the white dwarf. It piles up, thus slowly changing the physical condition inside the accretion disk. Eventually, at some point in the accretion disk, pressure and temperature become adequate for hydrogen to ionize. Neighbouring areas are "infected," and thus a wave of ionization starts travelling both inward, towards the white dwarf, and outward through the disk. Again the accretion disk is not able to adequately adjust to these new conditions, and the effect again is a collapse, with major accretion onto the white dwarf until the density and temperature of material in the disk again fall below the level where the ionization of hydrogen cannot be maintained; the disk cools off and the outbursts comes to an end.

Standstills are understood as prolonged accretion onto the white dwarf, which of course implies some kind of continuous supply of material from the secondary star for a while. Superoutbursts seem to be triggered by normal outbursts, but seem to be due, at least partly, to some different kind of process.

It is not clear, from a theoretical point of view, which of these models is correct. They both have their advantage, i.e. features that they can explain well, and their weaknesses; maybe some combination of the two comes closer to the "truth", or maybe something very different, which has not yet been deduced. For a number of observable features, these models make rather different predictions, which in principle can be tested and contrasted with each other. I'll discuss them in the following section.

2.2 Predictions

During the quiescent state, between outbursts, from the disk instability model it is to be expected that at least for a little while before the onset of an outburst, the optical flux would be seen to increase as the material accumulates at the outer edge of the disk and slowly heats up. The transfer instability model, on the other hand, would not lead to any changes in the accretion disk, or the disk might even be expected to cool a little as it is progressively being emptied. On the observational side, no clear distinction is possible, first because mostly the coverage of the quiescent part of the outburst light curve is not very good due to its low brightness and, secondly, because there are cases, like, in particular, SS Cyg, where in the same object all three possibilities (a flux increase, decrease, or no change between outbursts) are realized rather frequently, with no obvious relation to the shape of the outburst to come ... but this clearly is one point that deserves more thorough investigation.

The disk instability model predicts that the hump be essentially unchanged at all times, while in the transfer instability model there is the possibility that, shortly before the onset of an outburst, the hump could be seen to be considerably brighter than at other times, namely precisely during the (presumably short) period when the mass transfer from the secondary star is strongly enhanced. In spite of substantial observational effort during the past decade, mostly due to the unpredictability of the next outburst, such an enlarged hump could not yet be found with certainty in any system, nor could its definite absence be proven.

Due to the same effect, of whether or not an enhanced amount of material is pumped into the disk during onset, in the case of the transfer instability model the disk radius is expected to shrink temporarily (for reasons of conservation of angular momentum), while in the case of the disk instability model it clearly should remain unchanged. In some cases there seems to be some evidence that such a shrinking of the disk radius could be observed, but since it is very difficult to determine the outer radius, this issue is not clear, as simple temperature or pressure changes might cause the same effect.

Depending on what opacity[1] sources are used in the calculations of the dwarf nova outbursts, the disk instability model might predict a hold of several hours duration on the rise. No effect like this is expected in the transfer instability model. Again, the observational situation is totally unclear, partly for the unavailability of suitable observations, but also because existing observations have not yet been investigated exhaustively in order to settle this issue.

In the disk instability model, a difference in the change of the wavelength dependence of the emitted flux during rise is expected, depending on where in the disk conditions are reached first for the outburst to start, i.e. from where in the disk the instability fronts start to travel. If the outburst starts in the hot inner parts close to the white dwarf, which radiate both in the optical <u>and</u> in the UV, the optical and the ultraviolet flux are expected to rise simultaneously as the outburst spreads, while the rise should be seen first in the optical if the outburst starts in the cool outer parts, leading to the <u>UV-delay</u> or <u>UV-lag</u>. Also, the rise is expected to be rather fast ("normal") if it starts close to the white dwarf, while it should be slow ('anomalous') if it starts in the outer disk. In the case of the transfer instability model the ultraviolet delay always should occur, while the rise time always should be more or less the same. From the observational side this issue has not yet been settled. The UV lag was first detected in VW Hyi, an object where this effect is very pronounced, typically amounting to some 10 hours between V and the Lyman alpha area (1200 A). Fortunately, many observations of dwarf novae were taken on the (optical) rise with the IUE (ultraviolet sensitive) satellite over the years. For many of them it was possible long after the observations to, with the help of amateur observations of outburst light curves, determine that they actually were taken on the rise. From inspection of these it appears that probably the UV lag is a universal feature of dwarf nova rises, but the issue is not yet clear. One difference between different dwarf novae seems to be the duration of the lag. In some cases the lag is very short, in others it is very long and pronounced. There is an indication that the duration of the UV lag changes even within the same object from rise to rise. Also there appears to be some indication that fast rises in the optical mean strong UV lag, while in slow ones the flux rises almost simultaneously at all wavelengths.

---

[1]Opacity is a number describing the ability of atoms and molecules to absorb radiation of a certain wavelength; in particular at low temperatures reliable values are not always known.

No matter what the trigger for the rise was, decline should proceed in about the same way in both cases. Also, since it is expected to be basically a cooling, no distinctive structure in the outburst light curve is expected. In fact, however, a lot of structure is observed, but it is not clear what this is due to. This issue will be discussed later.

So much for predictions. In principle, a distinction between the two models should be possible, but for various reasons, could not be achieved yet.

2.3 Patterns

The outburst light curves should reflect a lot of the physics and workings of the accretion disk, and their careful study should help to reveal their secrets. Before the Roche model for the explanation of cataclysmic variables was found, i.e. at the time when even for professionals, short-term variations in dwarf novae were undetectable due to insufficiently developed technology, outburst light curves were just about the only information that could be obtained from cataclysmic variables, and consequently, they were the focus of research. Since then, the areas of interest have shifted, but nevertheless, light curves are not "out". The phenomenology of dwarf nova light curves has been investigated for many decades, and in the following I want to present some of the main findings. It is interesting to notice that theory and its predictions on one hand, and observations and patterns found in them on the other, do not yet go together. One or other of the observed features can be understood or, rather, makes some sense in the framework of the Roche model; others do not, and in general, the two areas of research have only just started to be coordinated.

All dwarf novae that I am aware of have a clear bi-modal distribution of the duration of outbursts, long ones and short ones. In some objects these alternate rather well; in others they occur more at random. There is some tendency for more long outbursts per short ones to occur the shorter the outburst period of a system, though averaged over longer intervals of time there never seem to be more long outbursts than short ones. The occurrence of outbursts of different lengths can be understood easily in the disk instability model, as the consequence of changed conditions in the accretion disk after each outburst; in the transfer instability model, theoretical difficulties with explaining this phenomenon are larger, but probably solvable. It should be stressed here that superoutbursts are different phenomena from long outbursts as several other features occur only in the context of superoutbursts, not during normal long outbursts.

The maximum brightness in most objects is more or less the same for all outbursts, though on the average the long outbursts tend to be somewhat brighter. Also, occasionally outbursts with very low amplitude occur in some objects, or even a complete cessation of outburst activity for some time, after which outbursts are resumed with apparently unchanged characteristics. Both these observations point to there being a reservoir of energy somewhere in the system which, when overfilled, gets emptied a bit more vigorously than at other times, while occasionally it is filled much more slowly than normal. The reason for the occurrence of these mini-outbursts is not clear.

The outburst periods are not periods in the strict sense of the word, but rather averages over long intervals of time, from which any particular value can deviate rather appreciably. This implies that there is some kind of a clock in the system, but obviously more that one determining physical parameter. Z Cam stars tend to have considerably shorter outburst periods than other dwarf novae. And in SU UMa stars, the periods of the superoutbursts follow a different, and much better kept, periodicity than the normal ones; thus the two kinds of outbursts seem to follow different clocks. Still, there is some evidence that, at least at times, normal outbursts are the trigger for superoutbursts. Also in superoutbursts, occasional distinct jumps in the mean period are observed: for many cycles the mean period has a certain value which is kept with a rather strict periodicity, then suddenly another not-very-different and equally-well-kept period is adopted; for each object, three or four such "fixed" values seem to exist between which the system switches at random.

Investigations in a few objects of the duration of quiescent intervals and slight changes of the mean brightness during quiescence have revealed that apparently both follow the same periodicity of some years (different for each object), which is strikingly similar to the activity cycles observed in cool stars. Thus the understanding is that these variations in cataclysmic variables have something to do with the activity cycle of the second stars.

Also investigations were made of the relation between the duration of the quiescent phase and the duration and/or brightness of both the preceding and the following outburst. In some systems, there is such a relation, though vague, with the preceding, in some with the following, in some with both, and in again others with none, of the outbursts; thus the picture is not clear at all. In general, a tendency seems to exist for longer outbursts to be preceded by somewhat longer quiescent intervals. The famous Kukarkin-Parenago relation which links the amplitude of outbursts with the average recurrence time is mostly based on combining these values for dwarf novae and recurrent novae. While for dwarf novae alone there is not much evidence for this relation to hold, if the (currently much favoured) understanding is correct that recurrent novae are powered by thermonuclear runaways, while dwarf novae are powered by some events in the accretion disk, it is not at all obvious that the two phenomena can reasonably be combined in order to derive some sort of universal relation.

There are still several other investigations and claims for relations between this and that observed feature to have been found, but evidence for these to be true is even less certain than for the classes presented above.

2.4 Bringing Theory and Observations Together

In summary, it probably would have to be stated that although there is abundant data, and although research in this area has been going on for many decades, and for about one decade supported by large and powerful computers, the gap between what is seen and what is theoretically understood is still large. The main reason for this is that it is a problem just at the limit of the capability of modern computers to compute outburst light curves of dwarf novae. No matter what the adopted model is, an outburst light curve can only be put together from a sequence of computed accretion disks, and already to compute these is very difficult. The main problem is that the nature of the friction between the particles in the accretion disk, the viscosity, is not well known yet, but at the same time it is this viscosity which governs more or less the entire outburst behaviour of an accretion disk. Also, the large range of conditions in an accretion disk makes it necessary to compute conditions in many places in the disk, and have them interact with each other. Once the development of a disk is computed, the radiation from each individual model is derived and thus the light curve over a longer time span is constructed.

On the part of the observations, it is necessary to extract from the wealth of different individual features in the light curves those that are characteristic or repeat frequently in each object, as well as in dwarf novae in general. Thus an "average" light curve can be constructed which contains all the important features that a model ought to reproduce.

In collaboration with the theory group in Tübingen (Prof. Ruder and co-workers), J. Mattei and I (on the observational side) are working on pursuing the path of research described above. Both the theoretical and the observational aspects of this work are still in progress, but here I want to present some first results that we obtained from the analysis of the best-observed dwarf nova, SS Cyg.

As a basis for our research, we use AAVSO observations from the years 1967 through 1987 which cover almost 200 individual outbursts. Through parametrisation of the large range of different features seen, some common structure and patterns in the light curve become apparent. So far we mainly have concentrated on the outburst parts of the light curves.

- Almost all rise and decline sections (in magnitude presentation) can be approximated satisfactorily with a string of straight line segments.

- A maximum of three segments per rise or decline is necessary.

- The structure of the rise and of the decline part of an outburst are independent from each other.

- Rise and decline parts of the outburst light curves are about equally complicated in structure.

- The more complicated a structure, the less frequently it is realized.

- Considering changes between negative and positive slope, all combinations are being realized that are possible with three independent segments.

- Some outbursts have such a complicated shape that no proper fit is possible.

- There is no obvious relation between the shape of rise or decline and the duration of the entire outburst.

As this is still unfinished research, no further results can be given here, but they will be published elsewhere in due course.

3. Orbital Light Curves

The analysis of outburst light curves can help resolve many questions about dwarf novae, but this, by the very nature of the observations, is limited to long-term effects. Changes in the structure of the accretion disk between and during outburst cannot be observed directly, for this a much higher time resolution is necessary. Long-term monitoring with high time resolution was tried by professional astronomers on a few occasions in the past, which yielded valuable results that demonstrate that this kind of observation provides the opportunity to monitor the accretion disk closely in its changes during the outburst cycle. And in this way, it is hoped that more clues can be obtained as to what the outbursts are really due to.

The difficulty for professionals, however, as in the case of monitoring outbursts, is that they rarely have telescopes available for more than a few night per year, so here again is an area where amateurs can contribute immensely to research.

Thus, I want to conclude with an appeal to amateur observers, to try to observe dwarf novae at orbital resolution. It seems that this kind of observation is just about at the current limit of what can be done with current amateur equipment, and, maybe, in the future, it will be even more easily possible. What is needed is a time resolution of some 5 min or better for, at best, one full orbital period (i.e. several hours) or more, per night, for many (if possible consecutive) nights. Dwarf novae should be monitored at all stages of activity, but in particular during the quiescent phase and during the, from this point of view most critical, rise to an outburst. The typical brightness in quiescence is magnitude 14 or fainter, and some 3 to 5 magnitudes brighter at maximum. Observations could be in white light, although of course more information could be gained if filters were used. The main problems I foresee are the weather, since only very good conditions would yield good-quality observations over a span of several hours; and that only rather large telescopes would be powerful enough to collect sufficient light. Clearly photoelectric equipment would be needed, and, finally, the observing procedure would have to be rather different from the way amateurs normally observe. The kind of observing mode needed, in order to gain the appropriately dense coverage of the flux changes, would be some sort of "high speed mode". In professional astronomy, this term is used for observations with time resolutions of seconds. This clearly is not what I have in mind, but the general way of performing such observations is the same: A comparison star is selected that at earlier occasions was confirmed to not be a variable; then sky and comparison are measured for a short time, while most of the observing time is spent on the object. Measurements are made every few minutes, but repeatedly, without going back to the sky and comparison very often. Only after some time, depending on weather conditions, maybe every 15 to 25 minutes, sky and comparison are measured. Then again the object is measured for a long time. For the reduction, the measurements of sky and comparison that were taken during the night are interpolated to provide the current value at the time of observation of the target.

4. Conclusion

For as long as dwarf novae have been observed, amateur astronomers, by carefully monitoring the outburst behaviour of these objects, have contributed an invaluable source of information to research. Such observations cannot practically be obtained by professional astronomers, simply because telescopes in general are not easily accessible to them. The beneficial contributions range from scheduling of observations to the observational basis for research on the dwarf nova outburst mechanism. The sort of observations almost exclusively used so far are observations of the outburst light curves. With the availability of ever better equipment, it is suggested that observations of orbital light variations in dwarf novae might be coming into the reach of amateur observations. If this could be done, a whole new area of research would be opened up.

Discussion

Walker: There is a problem with amateurs undertaking the kind of observations which you would like. A 0.5m telescope on a magnitude 15 star with white light will give about 100 counts per second. Dark current from the photomultiplier tube will be 10 to 100 counts per second. Therefore the observations are difficult, even with a 0.5m telescope. In addition, continuous monitoring of the variable, with only very infrequent references to the sky and the comparison star is dangerous in any but the very best sites in the world, or requires comparative, multi-object photometers.

Kilkenny: You can do the kind of observations which you suggest, in terms of "high speed photometry" - this has been done frequently at the South African Astronomical Observatory - but I agree with Norman Walker that a good site is necessary. You need a dark site with reliably constant atmospheric extinction.

COORDINATION OF MULTIWAVELENGTH GROUNDBASED AND SATELLITE OBSERVATIONS OF NOVAE IN OUTBURST

Steven N. Shore
GHRS Science Team/CSC, Code 681
Goddard Space Flight Center
Greenbelt, MD 20771, USA.

I have become convinced that the <u>whole</u> nova phenomenon must be studied; the variations of total light and continuum, of radial velocity, and of the intensities and profiles of absorption and emission lines must be seen as connected parts of one physical phenomenon, rather than as isolated data that can be understood separately. C. Payne Gaposchkin (1957)

1. Introduction

Cosmical objects show a wide range of behavior, both dynamical and photometric, but few are as rich or as maddening in diversity as novae. Intrinsically unpredictable, often faint, and generally quite short-lived, they call for rapid mobilization of observational resources in order to capture their fleeting phases. This review is intended to provide some justification for why novae are such engrossing physical laboratories, what the research of the past few years has revealed about the causes and development of the outbursts, and how the multi-wavelength approach to observations of novae, through the advent of new techniques and technologies, has opened up a wider scope of physical processes to detailed study.

The classic monograph on novae by Payne Gaposchkin (1957), from which the dedicatory quote above is derived, is still an unsurpassed source of historical data on the brightest novae observed in the last century. This book is a monument to optical observations, and summarizes virtually the entire research program until the beginning of space observations at the start of the 1980s. Duerbeck (1987) has produced a catalogue of galactic novae; it is a comprehensive bibliographical resource, and includes finder charts for all of the objects. The IAU circulars, the *Information Bulletin on Variable Stars*, and the AAVSO publications are invaluable records of the *real time* information concerning the outbursts. Most of the literature on novae is, unfortunately, quite scattered, often appearing in conferences on cataclysmic variables or in specialized meetings on observations at specific wavelengths. Only a fraction of the available data has been published in journals, rendering the summary of the wealth of data difficult. It is lamentable that, to date, there has not been an effective synthesis of all of the observational material now available that rivals *The Galactic Novae*. This review is thus intended to be a herald of things yet to be. The reader should note that, because of my prejudices, I will concentrate on the information gained from ultraviolet spectroscopy. Reference is made to reviews that will point the reader to a more dispersed literature of original studies.

2. A Brief Taxonomy of Novae

There are several observational classes of novae. *Classical* novae, like GK Per, DQ Her, or V603 Aql, generally consist of low mass white dwarfs, about 1 $M_\odot$, in very close binaries having periods less than about 12 hours. The companions are usually M dwarfs, also low mass objects. *Recurrent* novae, like T CrB, RS Oph, or V394 CrA, are a more diverse group. The term "recurrent" comes from the fact that they display repeated outbursts within the lifetime of the average observer, or at least within about one century. Some, like T Pyx, U Sco, and N LMC 1990 #2, resemble classical novae. They have low mass companions, although these may be stripped evolved stars (as evidenced from the high helium abundances observed in the ejecta from these systems). Others, notably RS Oph and T CrB, have M giant companions and orbit periods of hundreds of days, more like symbiotic stars. We will return to this class later in the review. Dwarf novae are an observationally distinct type of cataclysmic. Although many of their properties are similar to novae, dwarf nova systems do not display explosive ejections and will not be further discussed here.

An additional, although not independent, subdivision is based on the speed class, defined using the parameter t_3, the timescale required for decline of the nova by 3 magnitudes from optical maximum light (Duerbeck 1989). Fast novae, like U Sco (also a recurrent nova) and V1500 Cyg, have t_3 of order one week. Slow novae, like DQ Her, DK Lac, and DN Gem, have t_3 of many months. Some extreme slow novae, for instance RR Tel, may stay in an active state for years and look far more like symbiotics than classical novae. A few, notably η Car, have been reclassified on the basis of additional work as luminous blue variables or into other categories.

Finally, novae also can be separated on the basis of heavy element abundances. This is a more model-dependent scheme, based on line strengths in their spectra, especially in the later stages of the outburst when the ejecta turn optically thin. Normal novae show enhancements of the CNO group. A recent development is the discovery of the "neon nova" subgroup, novae displaying enhanced abundances of Mg, Al, and especially Ne. These novae are the product of massive oxygen-rich white dwarfs, near the Chandrasekhar mass limit. The CNO group of elements, especially nitrogen, appears to be the most systematically altered by the outburst nucleosynthesis. Depending on the abundances in the accreted material, which is occasionally helium rich if it comes off an evolved companion, nitrogen is systematically enhanced in many explosions by as much as two orders of magnitude over the solar abundance. Little is known at the moment concerning isotopic variations, but molecular observations suggest that ^{13}C may be enhanced over the solar value, again the result of incomplete CNO processing in an explosive environment.

A few novae have formed large dust envelopes late in the outburst, notably DQ Her and QV Vul (N Vul 1987) (Gehrz 1988). These are especially easy to distinguish on the basis of their optical light curves, which often display dramatic magnitude swings during which the star fades by several magnitudes and then slowly recovers on a timescale of months.

Let me say at the outset of this brief review that there is essentially no difference in the physical process that gives rise to various types of novae. The diversity arises in the initial conditions of the particular objects, the mass of the accreter, the nature of the companion, the mechanism for the accretion, whether the accreter has a magnetic field or not, the abundances present in the accreted material, and so on. But by their diversity, we are able to sample the full panoply of the physics underlying the nova phenomenon and it is for that reason that the more we know about individual novae and their histories, the more we understand about the unified mechanism that produces this dramatic cosmic event.

3. The Theoretical Framework

Novae occur in close binary systems in which one of the stars is degenerate (Starrfield 1986). This is essential to the physical understanding of the phenomenon. Regardless of the nature of the companion, with which we shall deal later, the fact is that novae cannot occur in binaries unless some form of mass transfer takes place onto a white dwarf or neutron star. The canonical model is that as mass, transferred from the companion, falls onto the surface of a compact object, it compresses and heats up in a standing shock. The combination of opacity and the release of gravitational energy in the surface shock layer on the degenerate star causes the temperature in the layer to rise. The accreted material is compressed by continued accretion and its abundance altered through mixing between the underlying envelope of the white dwarf and the accreted material.

It is now that the degeneracy plays a central role. The fundamental feature of a degenerate gas is that its equation of state, the relation between pressure, density, and temperature, departs from the ideal gas law: the pressure is independent of the temperature. Thus, even though the temperature continues to rise in the layer, the pressure increases only as a result of the increase in the density from the added superficial mass. Eventually, as the temperature continues its rise, the layer reaches the temperature high enough for nuclear fusion to commence. The sudden onset of nucleosynthesis releases an enormous amount of energy, heating the burning layer and, by a positive feedback, further increasing the nuclear reaction rate. The luminosity rises steeply in the outer layers of the white dwarf as a result of this added energy. However, there is initially no structural re-adjustment because the pressure is insensitive to this rise in the temperature. In fact, *the density does not change as the layer fuses progressively more matter.*

The luminosity may increase so much that radiation pressure disrupts the mechanical stability of the outer layers of the white dwarf. This maximal luminosity, called the Eddington limit, is given by $L_{Edd}=4\pi GMc/\kappa \approx 3 \times 10^4 L_\odot M_\odot^{-1}$, where G is the gravitational constant, c is the speed of light, M is the stellar mass, and κ is the opacity per gram of stellar material. This corresponds to a bolometric absolute magnitude of about $-6.^m6$, so a nova in the LMC at maximum may be as bright as 11^m, making them relatively easy objects even at 55 kpc for photographic patrols. This radiative driving leads to the equivalent of a massive stellar wind that expands very rapidly away from the accreter. In addition, the temperature may rise above 10^8K, the level at which the gas changes its equation of state and begins to act like an ideal gas, leading to an explosion. These two events, which are not necessarily independent, contribute to a net outward acceleration of the outer layers of the accreter. The envelope begins to expand and brighten, and the surface temperature begins to fall. This all takes place in a matter of several thousand seconds, usually far too rapidly to be caught by observation. However, the subsequent events are observable, and it is these stages that we shall deal with in greater detail. Either way, the characteristic timescale for this initial dynamical stage in the outburst is of order a few days.

The input of nuclear-generated energy produces a steep rise in the stellar luminosity. Because the layers have not yet begun to expand and cool, this first pulse is almost entirely in the X-ray and ultraviolet. Once the layers begin to accelerate, the expansion throttles the energy release and the nova enters a stage of nearly constant total luminosity. However, the rapid envelope expansion means that the wavelength of the peak of the spectrum moves from the UV to the optical as the envelope cools and becomes initially more opaque. Thus the visible light output increases, and this is the characteristic observational signature of the outburst.

Whether driven by explosive processes or radiation pressure, the ejecta velocity is expected to be several times the escape velocity from the white dwarf, of order several thousand kilometers per second. In some recent cases, however, the observed velocity has been dramatically in excess of this, reaching over 10^4 km s^{-1} for N LMC 1990 #2.

4. Observational Diagnosis of the Outburst

The diagnosis of the initial moments of the nova event require rapid response and multiwavelength coverage. As an example of how one proceeds to incorporate different wavelength regions, I will use the two recent novae: N LMC 1990 #1, for which we have been able to obtain nearly simultaneous optical and UV coverage, and the galactic recurrent nova V745 Sco (for observational details, *c.f.* Rohlfs 1990). Only a handful of novae have been well observed at multiple spectral regions from the X-ray to the radio: V404 Cyg, RS Oph 1985, V1500 Cyg, U Sco 1979, N LMC 1990 #1 and 1990 #2, OS And, V394 CrA, and GQ Mus. This list is biased by the requirement that ultraviolet observations have been obtained in addition to optical and infrared and possibly radio. Only a few of these novae have been observed by X-ray satellites (see Bode & Evans 1990).

The optical maximum signals that the shell is being ejected. Because the opacity is lower longward of the Balmer discontinuity, at 3647Å, radiation can preferentially escape here. The rapid cooling of the outer layers causes several dramatic, and important, changes in the spectrum of the white dwarf. As the temperature falls below about 12000 K, strong absorption bands of Fe II appear in the spectrum. These serve to increase the opacity and form a sort of "iron curtain", masking the hot dwarf and shifting the peak of the radiation longward to about 2800Å. The nova initially fades in the UV and remains opaque for quite some time, usually days to months. Yet even from the meager data we have for V745 Sco 1989, for example, it is obvious that the heavily extincted UV is fading more slowly than the optical light curve. For N LMC 1990 #1 it is more dramatic - the nova had faded to complete invisibility for the IUE optical sensors ($V > 14.^m5$) but was still strong in the ultraviolet. For a classical nova, where the mass of the ejecta may be as large as $10^{-5} M_\odot$, the ejecta remain optically thick and the nova light curve plateaus for up to several months. For lower mass ejecta typical of recurrent novae (see below), more like $10^{-7} M_\odot$, this stage is severely curtailed and may last only a matter of days if it occurs at all. Radio observations show that there is a shift of the peak of the emission toward shorter wavelengths with time, as the ejecta become progressively more optically thin (Hjellming 1988).

As the ejecta expand, the P Cygni profiles narrow. This is because the outer layers become optically thin, and thus the fastest moving matter ceases to contribute to the absorption trough on the line profile. This is most clearly seen in the LMC 1990 #1 C IV λ1550 resonance line profile. Deeper layers, more slowly moving and coming from lower depths of the processed zone of the white dwarf, can remain hot and optically thick for quite some time. The length of time required for the P Cyg profiles to disappear is also a measure of the mass of the ejecta. However, there are several ways to make the P Cyg profile disappear, or never form. The simplest is for the envelope to be ejected asymmetrically so that the effective covering factor for the core is small. In this case, the optical depth of the shell is always small and coronal lines will be seen rapidly. An alternative is for the shell to fragment. While observed in many optical spectra of novae in outburst, UV observations are thus far equivocal on this point, lacking the required spectral resolution. The disappearance of P Cyg profiles on the UV resonance lines signals the transition to the optically thin *nebular* phase.

For a few novae, especially RS Oph, the effects of the environment are readily visible in the evolution of the UV line profiles (see fig. 1). The wind of the red giant companion is flash ionized by the UV pulse from the explosion and the low density wind first produces absorption lines seen against the broad emission lines of the optically thick ejecta and later narrow emission lines which persist as the ejecta fade from view.

As the optical depth of the shell decreases, far ultraviolet radiation from the hot core can leak out. This radiation ionizes the ejecta, and anything else in the environment, producing a strong and persistent emission line spectrum. The emission line strength grows as the ionization eats outward through the ejecta, eventually reaching the edge and thereafter the emission line strength decreases due to the expansion of the ejecta and the decrease in the recombination rate, the physical process producing the emission. Depending on the amount of mass in the ejected shell, the emission line maximum and the onset of the optically thin phase of the expansion may be very slow. But eventual the shell becomes completely optically thin and completely ionized at this stage, the so-called "nebular phase", it becomes possible to determine the abundances of the elements synthesized during the thermonuclear runaway and mixed into the ejecta in the first seconds to minutes of the outburst.

The peak of the incident radiation field shifts into the far UV, and the brightness of the nova in the $\lambda\lambda 1200-3300$Å region decreases. At quiescence, an old nova has a temperature as high as 10^6K and emits primarily far ultraviolet (FUV) and X-ray radiation. Presently, there is little data on the physical condition of novae in quiescence. We do not know how the final stages of relaxation from the explosion take place, how the core re-establishes equilibrium and accretion recommences and how the mass loser responds to the aftermath of the explosion. All are intriguing physical problems, and all are currently at the far edge of our observational capabilities. Unfortunately, before proceeding further into the theory, we require observational constraints. The problem is a complex one involving both nucleosynthesis and hydrodynamics and, if unfettered by empirical input, is too rich in adjustable parameters for an *a priori* approach to be suitable.

5. Satellite Observations: *IUE*

In the past decade, the field has been revolutionized by a telescope about the size of a typical backyard amateur instrument. The *International Ultraviolet Explorer* satellite, also known as *IUE*, is a 45 centimeter telescope, operating between $\lambda\lambda 1150$ and 3400Å. The satellite is in geosynchronous orbit, operating 24 hours per day in three eight-hour shifts, divided between NASA (2 shifts) and ESA (1 shift). Observations are all real-time format, with the observer sitting with a resident astronomer and telescope operator at a console, commanding the satellite's activities and making instant decisions based on the progress of the observations.

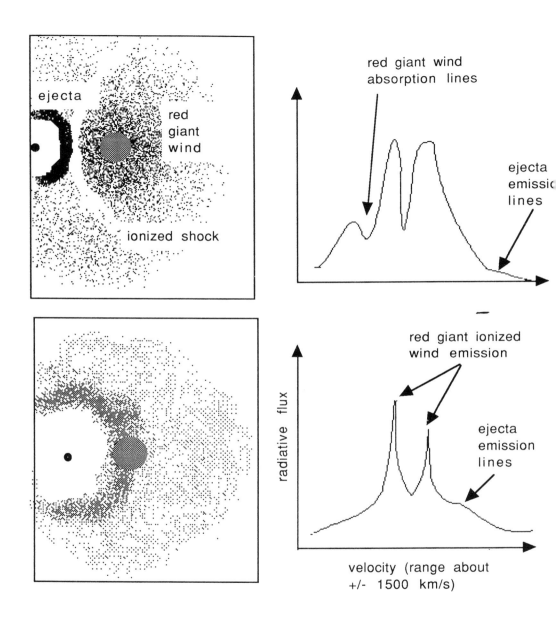

Fig. 1. Schematic of RS Oph 1985 in outburst. Similar phenomena were observed for V745 Sco 1989. The wind of the red giant is ionized and forms emission lines when the shell expands enough to become optically thin. Before that, the wind absorbs against the expanding ejecta.

The schedule and exposures can be altered at a few hours notice, if necessary, making IUE remarkable for its flexibility. The spectroscopic observations can be obtained in two dispersions, a low resolution mode with $R = \lambda/\Delta\lambda \approx 300$ and a high resolution mode with $R \approx 10000$. At high resolution, the nominal velocity resolution is about 30 km s^{-1}.

Novae are especially interesting objects for ultraviolet observations. First, during the outburst, most of the light emerges in the UV. Following the onset of the nebular phase, the strongest emission lines of many species, those best suited to abundance studies, are accessible only shortward of the atmospheric cutoff. They display an enormous range of ionization among the strong UV emission lines, *e.g.* He II $\lambda 1640$, C II $\lambda 1335$, C III] $\lambda 1910$, C IV $\lambda 1550$, N III] $\lambda 1750$, N IV] $\lambda 1490$, N V $\lambda 1240$, O III]$\lambda\lambda 1667$, 3133, Si III] $\lambda 1890$, Si IV $\lambda 1400$, and Al III $\lambda 1860$, the lines normally used for abundance studies. Finally, even though novae in quiescence are usually very faint at visible wavelengths, often having V > 16, they can still be observed in the UV. This is because they usually display temperatures in excess of 10^5K. Most of the historical novae are close to the Sun, usually within a kiloparsec or so, and they suffer low enough extinction to render UV observation possible even at minimum light.

The cooperation between groundbased and satellite observers has been critical in the continuing effort to understand the nova phenomenon. First, to date no nova has been discovered in outburst through UV observations. Optical patrol observers, very often amateurs, notify the community through the IAU circulars, phone calls, or computer networks. The rapid expansion of electronic mail networks has revolutionized many areas of astronomy, though perhaps nowhere as dramatically as in the field of nova and supernova observations. Teams have been formed by US and European observers to respond rapidly, via *target of opportunity* programs, to these announcements.

Often it is a judgment call. For several novae, like V745 Sco 1989, UV observations were possible for only a few days. The interstellar extinction proved to be too great to allow for reasonable exposure times after about one week of the outburst. The same was true for N Sco 1989. On the other hand, the two novae in the LMC that blew in 1990 were followed for about three months each, long after they were fainter than about 15*th* magnitude, because the line of sight extinction to the LMC is only about 0.^{m}1. In these four cases, observers at Goddard and Vilspa were able to begin observations within about one day of the announcement of the discovery. For N LMC 1990 #1 and #2, we were able to obtain UV spectra that actually showed that the nova had not yet reached maximum light in the UV even though the optical light was on the decline.

6. What Have We Learned in the Past Decade?

What have we learned from these observations? First, and most important, in the UV we see the fastest material in the outburst. Since novae occur as "instantaneous" explosions, that is that the layer erupting is geometrically thin compared with the radius of the star, the imposed velocity field is very simple. There is a range of initial velocities going from the top to bottom of the zone. The imposed velocity structure has the same dependence on radius as the expansion of the Universe, $r(t) \sim v$, where r is the distance of the zone and v is the velocity of expansion, and is referred to as a "Hubble flow". In other words, the material ejected with the highest speed moves farthest in the available time. The UV lines are the strongest, and thus we can see this very fast-moving material. Excited state transitions, like $H\alpha$ or $H\beta$, sample only the deeper layers, which have slower expansion velocities. Thus, relying only on optical data gives a misleading picture of the total bulk kinetic energy of the ejecta. Furthermore, the luminosity of the nova at maximum is grossly underestimated in the optical. For instance, only about ten percent of the luminosity of N LMC 1990 #2 emerged longward of 3000Å at maximum. The total luminosity was above the Eddington limit for the white dwarf, indicating that radiation pressure played a substantial role in the ejection event.

Optical and IR observations have demonstrated that some novae have higher than solar abundances of comparatively exotic elements like neon and magnesium. This is corroborated by the UV data. The "neon novae" were in fact discovered first by IUE observations. The strengths of the [Ne IV] $\lambda\lambda 1602, 2422$ and [Ne V]$\lambda\lambda 1575, 2974, 3346$ lines indicate the substantial enhancement of this element relative to solar values. In the early spectra of N LMC 1990 #1, the strength of the Al III $\lambda 1860$ doublet was quite remarkable, and suggested that this was also a member of the Ne-nova group; this conjecture was confirmed later by optical spectra taken during the nebular phase.

The nebular stage of nova development is accompanied by a host of complex phenomena. First because the shell is optically thin, the ionizing radiation from the hot corefloods the ejecta, and the environment. The increase in the strengths of optical coronal lines ([Fe VII], [Ca IX], etc) give ample evidence for temperatures in excess of 10^6K for the now relaxing core. In the case of recurrent novae, especially the recent outbursts of V745 Sco, V3890 Sgr. and RS Oph, the coronal line phase appears at about the point at which the ejecta become optically thin in the He II Lyman continuum, shortward of 500Å. Similar lines are not observed in the UV, but in the IR they play an important role in determining the ionization state of the envelope. The coronal lines are forbidden transitions, and in part their appearance signals the transition to extremely low electron densities in the ejecta. As the recombination rate decreases due to the expansion, these lines freeze out and eventually disappear. Their first appearance is a measure of the total mass of the ejected material, their disappearance is a result of the mass and velocity (the rate of density decrease) in the ejecta. For any line formed primarily by recombination, the drop in the emission measure of a line depends on $\alpha(T)n_e^2 V$, where α is the recombination coefficient, n_e is the electron density, and V is the volume of the emitting material. The electron temperature may become quite high, depending on the ions primarily responsible for the electrons, and this is reflected in the excitation rate for the coronal lines.

For the infrared, the formation of dust during the earlier stages of the ejection produces an important signature at about the same time. The UV irradiation of the dust, which is an effective absorber shortward of 3000Å, produces a dramatic increase in the IR emissivity. The dust may also be accompanied by neutral forbidden lines, and recently emission from CO has been discovered. The importance of these observations is to connect the formation mechanism of dust with a violent ejection event, in spite of the extremely high temperatures indicated for the gas in the ejecta from the atomic emission lines. Such phenomena have also been noted in SN 1987A, where the shape of the emission lines has been seen as a probe of the distribution of dust in the envelope of the supernova. By rapid response to an outburst at many wavelengths, the formation of dust can be tied into the overall physical state of the ejecta. Novae are likely important sources for condensates, but their role has not yet been clarified; clearly more observational work is needed to sort through the individual details of particular outbursts to reveal the underlying mechanism. One of the best examples recently observed among novae is N LMC 1988, which was observed to form dust within a few months of the UV maximum. Because we know the distance to this nova, the flux variation can be put in absolute units, allowing the dust mass fraction of the ejecta to (someday) be determined.

7. Coda: The Shape of Things to Come

In closing, I must emphasize that the importance of continuing to observe novae at many wavelengths rests with their individuality. If all novae were alike, if they were frequent enough that their regularity would become obvious, we would clearly be able to close up shop. But at the moment, there are only a handful of novae for which there exists the kind of multiwavelength observations that have been outlined here.

In the next decade, it should be possible to make extensive use of space-based observatories to augment the work of groundbased observers (and theorists). The Hubble Space Telescope, in particular the Goddard High Resolution Spectrograph (GHRS) and the Faint Object Spectrograph (FOS), will provide extensive ultraviolet and visible wavelength coverage at intermediate and high resolution. However, IUE continues to be the mainstay of UV spectroscopy. ROSAT, now completing its all-sky survey, will provide additional information about the soft X-ray and extreme ultraviolet spectrum, extending the wavelength coverage and the work begun with the UV Spectrometer on Voyager; pointed observations will commence at the end of the survey on specific targets, including novae. The Infrared Space Observatory (ISO), now in development, has spectroscopic and photometric capabilities that are important for nebular and dust diagnostics of novae. The Gamma Ray Observatory (GRO) will soon be launched and will be able to study novae in quiescence and study the nucleosynthetic products in their envelopes. The EUV Explorer will be launched soon and will conduct an all sky photometric survey that will likely reveal information about the quiescent old nova state. Lyman-Far Ultraviolet Spectroscopic Explorer (FUSE), currently in design, will also be important for the study of novae in quiescence and outburst in the wavelength region 100 - 1200Å. Groundbased observatories continue to improve their high resolution IR spectroscopic capabilities, and develop progressively more sensitive and higher angular resolution infrared imaging instruments that will be vital for the study of the late stages of the ejecta expansion.

Amateur astronomers will play critically important roles in the years ahead. Without early notification of the optical outburst, UV, optical, IR, and radio observations cannot provide information about the earliest stages of the explosion and much, if not all, of the interesting hydrodynamical information is lost. Dedicated surveys are important for monitoring recurrent novae as well as for discovering first outbursts of new ones. It is not just enough to know that an outburst occurred some time ago. We need to be able to respond quickly so that we don't lose the chance to see the critical moments of the outburst during which the important physical processes manifest themselves. There is still much to do, and the tools on the horizon hold out spectacular promise for the next decade of nova research.

I wish especially to thank warmly my collaborators Sumner Starrfield, George Sonneborn, and Bob Williams. I also thank Bob Gehrz, Bob Hjellming, Mike Dopita, Angelo Cassatella, Joachim Krautter, Joanna Mikolajewska, and Scott Kenyon for discussions and their willingness to share their data and experience. Finally, I wish to thank Janet Mattei and John Percy for their invitation to this conference and NASA for continuing research support.

SELECT REFERENCES

Bode, M. F. (ed.) 1987, *RS Ophiuchi: The 1985 Outburst* (Manchester: Univ. of Manchester Press).
Bode, M. F. and Evans, A. (eds.) 1989, *Classical Novae* (NY: J. Wiley).
Duerbeck, H. W. 1987, Space Sci. Rev., 45, 1: "A Reference Catalogue and Atlas of Galactic Novae".
Gehrz, R. D. 1988, *ARAA.*, 26, 377. (Infrared observations of novae in outburst).
Hjellming, R. M. 1988, in *Galactic and Extragalactic Radio Astronomy: 2nd Edition* (eds. Verschuur, G. and Kellermann, K.) (Berlin: Springer-Verlag).

Payne-Gaposchkin, C. 1957, *The Galactic Novae* (NY: Dover).
Rohlfs, E. (ed.) 1990, *Evolution in Astrophysics: 12 Years of Ultraviolet Astronomy with IUE* (ESA-SP) *in press*.
Starrfield, S. 1986, in *New Insights in Astrophysics* (ESA SP-263) (ed. E. Rohlfs) p. 239.
Starrfield, S. and Snijders, M. A. J. 1989, in *Exploring the Universe with the IUE Satellite* (ed. Y. Kondo) (Dordrecht: Kluwer), p. 377: "Galactic Novae".

IUE OBSERVATIONS OF MASS-LOSING CATACLYSMIC VARIABLES

Janet E. Drew
Department of Physics
Keble Road
Oxford OX1 3RH, England

1. Introduction

The realisation that mass loss occurs in association with mass accretion in cataclysmic variables (CVs) followed upon early observations made by the "International Ultraviolet Explorer" satellite (IUE) (e.g. Krautter et al. 1981). The types of CV we know are capable of driving winds are those with complete accretion disks - the nova-like variables and dwarf novae. It also seems that the mass accretion rate has to be high because dwarf novae probably only lose mass when they are in outburst. The kinds of questions that we would like to answer about these winds are the sort: "how much mass is lost?", "where do the winds originate?" and "how are they powered?". As answers are found, we will be better able to assess the state of the disks and white dwarfs in these systems and will know if the mass loss has any evolutionary implications. CV winds may affect long-term evolution in one of two ways: (i) by limiting the effective rate of mass transfer or, (ii) by providing a means of angular momentum loss.

In this article I aim to summarise what is known about CV winds already and to describe a continuing programme of IUE observations that has benefitted greatly from input from the amateur community of observers. Fuller reviews of CV winds may be found in papers by Cordova and Howarth (1987), Drew (1990, 1991) and LaDous (1990).

2. A Summary of Present Understanding

2.1 Wind mass loss rates

The fact that there is mass loss from nova-like variables and dwarf novae is made known by the presence of broad blueshifted absorption features in the ultraviolet spectra of these objects. This discovery had to await ultraviolet observations of these objects because it is only the well-populated ground states of highly-ionised heavier elements that are capable of producing the tell-tale absorption (transitions from these levels happen to fall in the ultraviolet, while transitions at optical wavelengths typically arise from sparsely-populated excited levels that do not provide large enough columns of absorbers). Figure 1 shows an example of the UV spectrum of the nova-like variable RW Sex, in which the wind-produced absorption features are particularly well-developed. The widths of the observed broad-absorption feature set lower limits on the wind expansion velocity: these are found to be in the range 3000-5000 km s^{-1}.

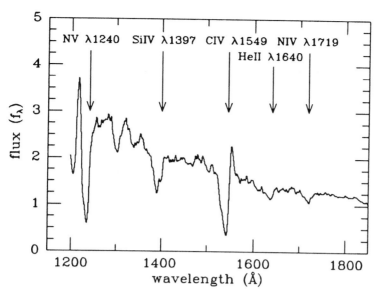

Figure 1. *A mean short-wavelength spectrum of the nova-like variable RW Sex constucted from image numbers SWP 16269-16271. Note the strong, broad wind-formed absorptions in the transitions N V λ1240, C IV λ1549 and Si IV λ1397. Wind absorption may also be present in the lines He II λ1640 and N IV λ1719. These data were extracted from the IUE Uniform Low Dispersion Archive (Wamsteker et al. 1989)*

The UV spectra of eclipsing CV's undergoing mass loss have a quite different characteristic form. Instead of broad absorption features, strong broad emission lines are observed. The clue that these emission lines are also produced in winds is given by the observation that the line eclipses are very much shallower than the eclipses of the continuum: this shows that the line-forming region is very much more extended than the region forming the continuum (i.e. the disk). Emission rather than absorption is observed because the change in viewing angle means that the bulk of the line-forming region in the wind does not shadow the smaller apparent area of the continuum-emitting disk. Figure 2 shows an example of an eclipsing system's UV spectrum.

To derive mass loss rates from the observed wind absorption or (in eclipsing systems) emission it is necessary first of all to compare observations with theoretically synthesised line profiles and then to estimate a correction for the abundance relative to hydrogen of the species responsible for the line. This procedure has now been applied to a number of objects. For example, Mauche and Raymond (1987) obtained $\dot{M}\xi \sim 10^{-11}$ $M_\odot$ yr^{-1} for HL CMa in outburst using the C IV λ1549 line profile. Drew (1987) estimated $\dot{M}\xi \geq 10^{-10}$ $M_\odot$ yr^{-1} for the eclipsing nova-like variables RW Tri and UX UMa using these objects' C IV emission. The quantity ξ is the fraction of all carbon present as the C IV ion. Its value is as yet poorly constrained, but is may be as little as 10^{-3} which in turn implies wind mass loss rates of the order of 10^{-8} $M_\odot$ yr^{-1}. At this level, the mass accretion and mass loss rates are comparable!

Table 1: Objects studied for line profile variability with IUE

Object name	Object type	Variability? detected?	Variability? orbital phase linked?
YZ Cnc	dwarf nova (SU UMa)	yes	yes
SU UMa	dwarf nova (SU UMa)	yes	yes
IR Gem	dwarf nova (SU UMa)	yes	yes?
RX And	dwarf nova (Z Cam)	slight	maybe
SS Cyg	dwarf nova (U Gem)	slight	maybe
DX And	dwarf nova (U Gem)	slight	maybe
V3885 Sgr	nova-like variable	yes	no
0623+71	nova-like variable	yes	no

2.2 The source of the outflow

The simple fact that velocities of up to ~5000 km s^{-1} are attained in these winds was quickly interpreted as indicative of mass loss originating from the vicinity of the white dwarf (Cordova & Mason 1982). This is because the escape speed from the surface of a white dwarf is of this order and is thus the magnitude of expansion velocity to be expected of an outflow initiated in its neighbourhood. Further evidence in favour of this view is provided by the failure to detect any blueshifted absorption in the UV spectra of eclipsing systems. If the outer disk contributed a significant amount to the outflow there would be shadowing of the UV continuum emission from the centre of the disk that, contrary to observation, would be apparent as absorption. It has yet to be clarified whether the outflow begins mainly in the inner disk or whether the accreted matter begins to settle on the white dwarf before it is ejected again.

Another implication of the lack of blueshifted absorption in eclipsing systems is that the outflow is likely to be stronger over the white dwarf poles than away form its equator (Drew 1987). In other words, these winds are likely to be somewhat bipolar rather than spherically-symmetric.

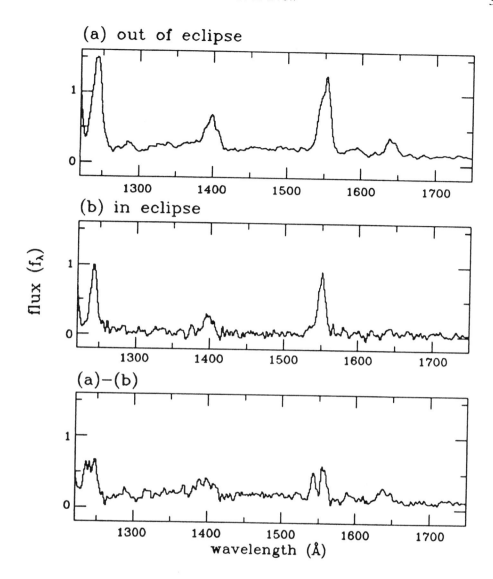

Figure 2. IUE spectra obtained during an outburst of the very deeply-eclipsing dwarf nova, OY Car: (a) mean spectrum out-of-eclipse, (b) mean spectrum during eclipse. The bottom panel contains the result of subtracting (a) from (b): this shows that the eclipsed C IV flux is distributed in a double-peaked profile, suggesting a disk component. OY Car is the only object in which this double-peaking has been noted to date. However it is typical insofar as the C IV, N V and Si IV line eclipses are much shallower than the continuum or He II $\lambda1640$ eclipses. These data were originally published by Naylor et al. (1988).

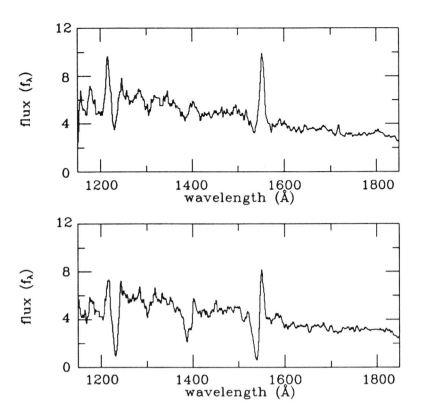

Figure 3. Two outburst spectra of the dwarf nova, YZ Cnc, obtained in the same IUE image (SWP 30237) half an hour apart. A full discussion of these observations is given by Drew and Verbunt (1988).

3. The IUE Observing Programme

The particular programme to be described began in 1987 when Frank Verbunt and I were awarded two ESA IUE shifts to observe the dwarf nova, YZ Cnc, during outburst. The goal of our observations was to find out if the variations in the UV wind lines apparent in archive spectra of YZ Cnc were linked to orbital phase. With a remarkable sense of timing, YZ Cnc went into outburst after a much larger than usual interval on March 5 1987: had the outburst not been relatively delayed, it is most likely that the observations that were in fact made on March 6 and 7 would not have been made until many months later because of the pressure to keep IUE pointing towards SN 1987A (the ultimate target of-opportunity!). The spectra obtained of YZ Cnc turned out to contain quite dramatic confirmation of our original suspicion: strong line profile variability was present and the timescale for variation was consistent with YZ Cnc's orbital period (2.086 hours, Shafter & Hessmann 1988). This behaviour is all the more surprising in view of YZ Cnc's low

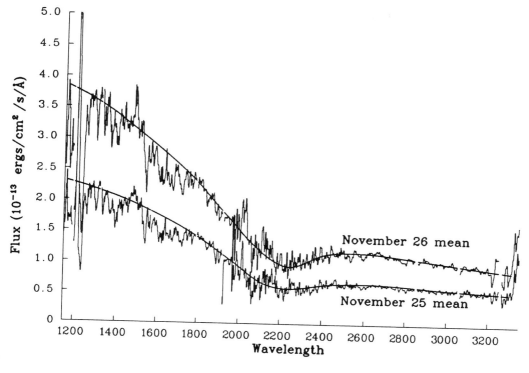

Figure 4. Merged short and long-wavelength mean spectra for DX and obtained on November 25 and 26 1989. The solid lines superimposed on the spectra are spline-fit continua. Of special note are the broad dips in the continua centred on $\lambda \sim 2200$ Å: these are caused by interstellar reddening.

inclination to the line of sight ($i < 50°$). Figure 3 contains two spectra obtained on March 6 in quick succession. In the earlier of the two only weak blueshifted absorption is apparent, while in the spectrum obtained about 30 minutes later the absorption has become very deep indeed.

These observations have since been followed up with an effectively long-term programme of IUE observations aimed at establishing whether any other CVs with winds show similar variations. Those involved with this programme are Frank Verbunt (Utrecht), John Woods (Sussex) and myself. The total allocation of IUE shifts we have received to date has been 21 (8 have yet to be used). The objects that have been observed as part of this programme are listed in Table 1. Significant line profile variations have been found in all but two of the objects observed, but only in the cases of SU UMa (Woods et al. 1990) and IR Gem (unpublished) are we confident that the variations repeat on the orbital timescale. It is tempting to guess that it is no coincidence that the systems behaving in the same way as YZ Cnc are also dwarf novae that undergo superoutbursts in addition to normal outbursts. Further observations are required before correlations of this sort can be substantiated.

Before going to discuss observations of IR Gem and DX And specifically, some remarks concerning the significance of line profile variability are appropriate. Variations linked to orbital phase are a firm indication of departures from axial symmetry either in the continuum-emitting disk or in the wind shadowing the disk. This follows from the fact that our sight line into a particular system is inclined, regardless of orbital phase, at a fixed angle with respect to the disk rotation axis (at low inclination angles to the orbital plane and in UV light, it is safe to leave the cool secondary star out of consideration). The problem is now to decide whether to locate the asymmetry in the disk or wind in systems such as YZ Cnc. Irregular variation of the kind seen in objects like V3885 Sgr probably has a quite different origin: it is likely to be due either to fluctuations in the wind mass loss rate or to wind ionisation changes. The fact that these variations can occur without changes in the ultraviolet continuum level suggests, quite plausibly, that the as-yet unobserved EUV radiation from on and close to the white dwarf surface controls the wind's state.

3.1. Observations of IR Gem

The way in which we came to observe IR Gem illustrates very nicely the value of rapid communication among amateur CV observers and between professionals and the central contacts provided by AAVSO headquarters and other individuals such as, in this case, Guy Hurst of 'The Astronomer' in the United Kingdom.

We had decided that in the absence of any more 'deserving' dwarf nova outbursts that our allocation of two IUE shifts in April 1990 would be used to observe V3885 Sgr, the nova-like variable. The timing of the IUE shifts within the day were such that it was especially advantageous to gather reports from European amateur observers via Guy Hurst. On the basis of a notification of a superoutburst of IR Gem and a confirmation within the same evening, we were able to switch the second of the two IUE shifts to a series of observations of IR Gem in place of V3885 Sgr. Prior to this, IR Gem had not been on the list of likely targets for our variability study because the probability of catching it in a bright enough state to allow short exposures is low. Hence the opportunity to make what were the first outburst UV observations of this dwarf nova only arose because of what was, in effect, an unsolicited alert.

The results of the UV observations have not yet been prepared for publication and so only qualitative comments about the data can be made here. In short, IR Gem appear to have ultraviolet wind-formed line profiles during outburst that are almost as variable a those seen in YZ Cnc. The timing of the sequence of short-wavelength exposures obtaine by Frank Verbunt was such that excellent phase coverage was achieved in a single shift (in effect a complete orbit was sampled at a mean phase interval of 0.15). Casual inspection of the data suggests that the pattern of variation as well as its amplitude is very like that seen in YZ Cnc.

3.2 Observations of DX And

As in the case of IR Gem, our observations of the dwarf nova DX And were made only because this system's outburst state became known to us via our contact with amateur observers. During our IUE observing run in the latter half of November 1989, Janet Mattei told me about the outburst and its progress in our regular phone contacts. It has to be admitted that if Janet had not told me, we would not have noticed the IAU circulars on the subject! As a result of this a decision was made to use the last two of five scheduled shifts (November 25 and 26) to observe DX And. These were the first ultraviolet observations to be made of this object which, in our view, has been surprisingly little-studied at any wavelength. The results of our study have recently been submitted for publication.

At the time of our observations, DX And was completing its rise to outburst maximum at $m_v \sim 12.1$. On November 25 the flux levels both in the ultraviolet and at optical wavelengths was 0.6 of those on November 26. It was immediately apparent after merging mean short and long wavelength spectra from each IUE shift that the reddening towards DX And is significant. The merged mean spectra are shown as Figure 4: the presence of the broad ~2200 Å dust absorption feature is very clear. From these data we have determined a colour excess of $E(B-V) = 0.20 \pm 0.04$. This in itself shows DX And to be somewhat unusual. The reddening sets a very loose lower limit of 400 pc upon the distance to DX And.

Furthermore, a low dispersion optical spectrum obtained by Bruch (1989) while DX And was quiescent is exceptionally red for a dwarf nova. This is an indication that the companion star is significantly more massive or more highly-evolved than the norm. It is also in keeping with this interpretation that the mean interval between dwarf nova outbursts is relatively long (~11 months). Our IUE observations provide a surprising clue to DX And's presumably long orbital period: radial velocity variations of the wind-formed C IV $\lambda1549$ and N V $\lambda1240$ show signs of a periodicity of the order of 11 hours. In November 1990, John Woods and I will attempt to obtain a more certain period determination from the time-resolved optical spectroscopy in La Palma. Even though the UV resonance line profiles do show velocity shifts there is little short-term variability in the profile shapes (if there had been, we would not have been able to measure velocity shifts at all reliably!).

On the basis of our existing IUE dataset, we have obtained an allocation of three further IUE shifts to be taken up by the middle of 1991. We aim to use these to obtain a quiescent ultraviolet spectrum and, if DX And cooperates, to follow its rise to outburst upwards from $m_v \sim 13.5$. To achieve this latter aim, we shall again depend upon prompt notification of the beginning of the rise from amateur observers.

4. Concluding Remarks

A major aim of this article is to show how valuable our contacts with the community of amateur observers have been. All our ultraviolet observations of dwarf novae have depended crucially upon outburst alerts from amateurs. We have gained much from both solicited and unsolicited requested notifications. Our work and also that of other groups is continuing with IUE and we look forward to obtaining a much clearer picture of the origin of orbital-phase-linked profile variations and indeed of the wind phenomenon as a whole.

References

Bruch, A. 1989, A & AS, 78, 145.
Cordova, F.A. & Howarth, I.D. 1987, in 'Exploring the Universe with the IUE Satellite', eds. Y. Kondo et al., Reidel pp 395-426.
Cordova, F.A. & Mason, K.O. 1982, 260, 716.
Drew, J.E. 1987, MNRAS, 224, 595.
Drew, J.E. 1990, proc. IAU Coll. No.129, in Physics of Classical Novae, eds. A. Cassatella & R. Viotti, Springer-Verlag, Berlin.
Drew, J.E. 1991, proc. IAU Col. No.129, in Structure and emission properties of accretion disks, eds. C. Berthout, C. Collin-Souffrin & J-P. Lasota, Editions Frontiers, Paris.
Drew, J.E. & Verbunt, F. 1988, MNRAS, 234, 341.
LaDous, C. 1990, in 'Cataclysmic Variables and Related Objects', Ed., M. Hack, NASA/CNRS Monograph Series on Non-Thermal Phenomena in Stellar Atmospheres.
Krautter, J., Klare, G., Wolf, B., Duerbeck, H.W., Rahe, J., Vogt, N. & Wargau, W. 1981, A & A, 102, 337.
Mauche, C.W. & Raymond, J.C. 1987, ApJ, 323, 690.
Naylor, T., Bath, G.T., Charles, P.A., Assall, B.J.M., Sonneborn, G., van der Woerd, H. & van Paradijs, J., 1988, MNRAS, 231, 237.
Shafter, A.W. & Hessmann, F.V., 1988, AJ, 95, 178.
Wamsteker, W., Driessen, C.D., Munoz, J.R., Hassall, B.J.M., Pasian, F., Barylak, M., Russo, G., Egret, D., Murray, J., Talavera, A. & Heck, A. 1989, A & AS, 79, 1.
Woods, J.A., Drew, J.E. & Verbunt, F. 1990, MNRAS, 245, 323.

TT CRATERIS 1987 - 1990

Richard Fleet
60 Blacklands Drive,
Hayes End, Middlesex
UB4 8EX
England

The Observations

The light curve since discovery, Fig. 1, is presented in blocks, each covering a single observing season. Daily means or the faintest negative estimate are shown.

Eleven outbursts have been reported so far, of which two remained brighter than magnitude 14.2 for at least 3 days and a further six for between 6 and 9 days. Only one outburst had a duration less than 3 days. For two others the duration is unknown. Overbeek also suspected a further two outbursts but sky conditions prevented confirmation. Most outbursts reach magnitude 13.1 at maximum but the shorter one reached only 13.7 (if caught at maximum).

In the first and third seasons the longer/brighter outbursts have a mean cycle of about 100 days, but there was a 140-day gap between outbursts in the fourth season (assuming none were missed). Rather curious is the 60 day, followed by 17 day, gap between reported outbursts at the end of the second season. Taking the latter to be caused by the shorter outburst, the mean cycle of the longer outbursts would now seem to be 100 ± 40 days. Being close to the ecliptic means that interference from moonlight and gaps near conjunction make it rather difficult to establish mean cycles with any confidence. In any case it will take many years for the star to show its full range of behaviour.

Acknowledgements

Thanks to alerts by observers, extra observations, including a first photograph and two spectra, were obtained. Guy Hurst of The Astronomer has been particularly helpful in passing on this sort of information.

All the estimates have been adjusted to fit a new sequence. V and (B - V) values for five comparison stars were determined by Pam Kilmartin, Mt. John University Observatory, for the VSS RASNZ. I thank Frank Bateson for permission to use these values in advance of their publication in Publ. 16, VSS RASNZ.

I thank Frank Bateson for supplying observations on behalf of the VSS RASNZ and Janet Mattei for supplying observations on behalf of the AAVSO.

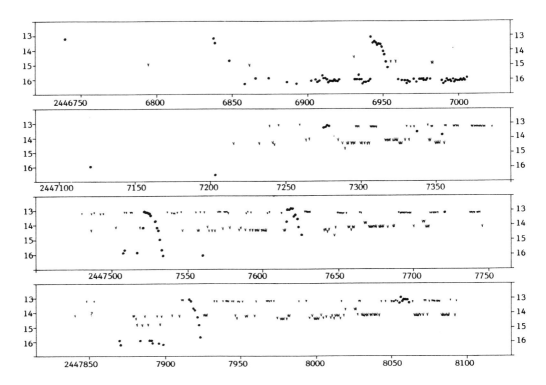

Thanks too to those observers, particularly Danie Overbeek, who supplied observations directly.

Although the latter half of the first season shows what can be achieved by a determined observer (in a good climate) no single person or group can hope to achieve complete coverage on an object like this. Having been unable to observe this object for most of the last three seasons I am especially grateful for the efforts of the observers listed in Table 1.

The whole exercise, particularly the support received, has given great satisfaction, no doubt enhanced by the fact that less than four years ago this object was just a query in my observing notes.

Table 1. Observer Totals, TT Crateris 1987-90.

Source	Observer	Total	Region
A	J.Bortle	5	U.S.A.
A	G.Dyck	13	U.S.A.
A	R.King	2	U.S.A.
O A	P.Sventek	10	U.S.A.
O	J.Abbott	2	EUROPE
O	R.Arbour	1	EUROPE
A	R.Fidrich	1	EUROPE
O	R.Fleet	15	EUROPE
O T	J.Toone	22	EUROPE
O T	W.Worraker	3	EUROPE
T	A.Young	1	EUROPE
O Z	M.Begbie	6	SOUTHERN AFRICA
O A Z	T.Cooper	38	SOUTHERN AFRICA
O A T Z	R.Fleet	138	SOUTHERN AFRICA
A Z	J.Hers	4	SOUTHERN AFRICA
O A Z	D.Overbeek	298	SOUTHERN AFRICA
O	J.Vincent	1	SOUTHERN AFRICA
A Z	T.Cragg	4	AUSTRALIA NEW ZEALAND
Z	W.Goltz	1	AUSTRALIA NEW ZEALAND
Z	A.Jones	52	AUSTRALIA NEW ZEALAND
T Z	A.Pearce	24	AUSTRALIA NEW ZEALAND
Z	B.Tregaskis	1	AUSTRALIA NEW ZEALAND
		642	

Source codes: O = Observer,
A = AAVSO,
T = The Astronomer,
Z = VSS, RASNZ

VISUAL DETECTION OF SUPERHUMPS IN SW URSAE MAJORIS

Bjorn H. Granslo
Institute of Theoretical Astrophysics
P.O. Box 1029, Blindern N0315 Oslo 3
Norway

1. Introduction

SW UMa is a long periodic dwarf nova. Outbursts are rare and only three have been detected since 1981. During its eruptions, the magnitude rises to magnitude 9 or 10 or about 7 magnitudes above the minimum level; the star remains above minimum light for about 25 days. The eruptions are characterized by a rapid initial rise followed by a slower decline. When the brightness has fallen to about 3 magnitudes below the maximum light a more rapid decline follows towards minimum. Such a behaviour resemble the supermaxima that are characteristic for the SU UMa subclass of dwarf novae. Supermaxima are bright and long eruptions that occur in addition to or instead of the normal outbursts. A study by Shafter, Szkody, and Thorstensen (1986) showed that SW UMa is a close binary system with an orbital period of 81.8 minutes. This period is also typical for SU UMa stars. During the outburst in 1986 March, Robinson et al. (1987) observed the star with a high-speed photometer at McDonald Observatory in Texas. The star was monitored on 10 nights using telescopes with apertures of 0.9 and 2.1 meters. During most of the time they detected superhumps with a period of 84.0 minutes and amplitudes of up to 11 percent. This proved that the eruption was a supermaximum and SW UMa is now classified as a SU UMa type dwarf nova. In addition to these large scale variations, small-amplitude oscillations with periods of about 5 minutes and flickering with mean periods of 22 seconds were detected on several nights. All subsequent comparisons to the 1986 outburst refer to the paper by Robinson et al. (1987).

The last outburst started on 1990 March 13 UT, nearly four years after the previous one. In several AAVSO Alert Notices (Mattei 1990a, 1990b) observers were encouraged to monitor the star closely as the superhump variations might be detected visually. The writer observed SW UMa during several nights and the results from 1990 March 21 and March 26 will be presented in this report.

2. Observations

The recent eruption of SW UMa lasted for more than 24 days as the star was still above minimum light on 1990 April 5. A light curve of the outburst is shown in figure 1 and it is based on observations from various sources, including IAU circulars and AAVSO Alert Notices. Hurst (1990) has communicated observations that define the final decline of the outburst.

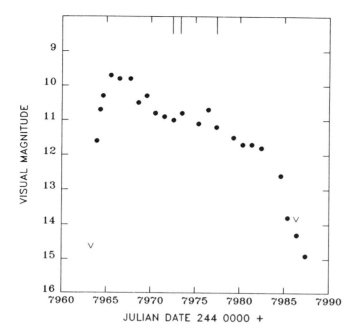

Figure 1: Visual light curve of the 1990 eruption of SW UMa. The results are presented as daily means except for observations on JD 244 7963 and 244 7964 where half day means were used.

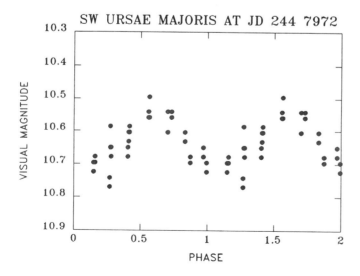

Figure 2: Visual light curve for 1990 March 21.

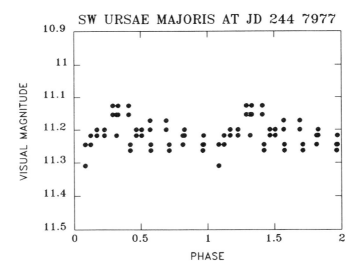

Figure 3: Visual light curve for 1990 March 26.

SW UMa was monitored on three nights in an attempt to detect the superhumps. The data from the first and third night will be presented. During these nights, the star was followed for about three hours, or two superhump cycles. The observations from March 22 covered only one cycle, as observing had to be stopped due to intervening clouds. The results are summarized in table 2. The ticks above the curve in figure 1 show when searches for superhumps were attempted.

Table 1: Visual monitoring of SW Ursae Majoris

UT Start 1990	UT End 1990	Duration of observing	Number of step estimates	Mean visual magnitude
d	d	d		m
Mar 21.866	Mar 21.991	0.125	32	10.65
Mar 22.810	Mar 22.865	0.055	16	10.70
Mar 26.827	Mar 26.947	0.120	38	11.20

As the amplitudes of the superhumps for SU UMa stars always have been reported to be of 0.3 magnitudes or smaller it was important to maximize the accuracy of the observations. The step method developed by F. W. Argelander (1799 - 1875) was used for all estimates. All observations were carried out using a 20 cm catadioptic telescope with a magnification of 80x.

Four comparison stars were used for the estimates and for the development of the step scale (grade numbers). These stars are identified in table 2. The magnitudes for star A and D were from the AAVSO chart for SW UMa (version 1990 March). These magnitudes appeared reasonable as compared to other comparison stars on the AAVSO chart. The step scale was deduced from the observations and the mean values are listed in the table.

Table 2: Comparison stars for SW Ursae Majoris

Comparison star	Delta RA m	Delta Dec '	Step value grades	AAVSO magnitude	Deduced magnitude
A	- 1.6	+ 8	0.0	9.9	9.90
B	- 1.7	- 8	6.0		10.45
C	- 0.4	- 27	10.7		10.88
D	- 1.1	- 17	16.4	11.4	11.40

The coordinates are relative to SW UMa (comparison star minus SW UMa). The deduced magnitudes were found by using equations (2).

Figures 2 and 3 show the light curves for March 21 and March 26 and the magnitudes have been plotted as a function of phase. The observations have been averaged over two cycles in order to reduce the effects of any systematic errors in the individual estimates. The curves from the 1986 outburst show that the shape of the light variations varies during the different cycles. However, when comparing two conclusive cycles the differences do not appear to be larger than that this procedure should be acceptable for visual observations.

The phase has been calculated from the equation

$$\text{Phase} = \text{FRAC} ((JD - 244\ 7972) / 0.0583) \qquad (1)$$

where FRAC denote the fractional part of the expression and JD is the Julian Date for the time of the observation. The number 0.0583 denote the superhump period in days.

Using the grade values in table 2 the brightness of the variable was calculated in step units for each estimate. The grade values (gv) were transformed to magnitudes (mv) by using the linear transformation

$$mv = 9.9 + 0.0915 \, gv \qquad (2)$$

The number 0.0915 was found by dividing the magnitude difference between star A and D by the corresponding step difference.

3. Discussion

The observations from March 21 (JD 244 7972) and March 26 (JD 244 7977) indicate a period of about 90 minutes which agrees well with that which was found during the 1986 eruption. Clear and regular variations were detected during the first night. The amplitude was nearly 0.20 magnitudes and considerably larger than the scatter in the light curve that amounted to about 0.10 magnitude. On March 26, the variations were smaller and more asymmetric but the light curve was more difficult to interpret as the amplitude appeared to be of the same order as the scatter. Observing was more difficult as the star was markedly fainter on this night. The characteristics of the superhump variations are presented in table 3.

Table 3: Characteristics of superhumps

Date 1990 UT	Range of magnitude	Amplitude in magnitudes	Phase of maximum	Phase of minimum	Time of rise in phase units
March 21	10.55-10.75	0.20	0.15	0.60	0.55
March 26	11.15-11.25	0.10	0.05	0.35	0.70

The observations from March 21 and March 26 were made 8 and 13 days after the onset of the eruption. The variations resembled the behaviour on 1986 March 13 (11 days after the onset) and 1986 March 16 respectively. It is interesting to note that superhumps were not present 6 days after the onset of the 1986 outburst but were clearly visible 8 days after the beginning of the 1990 eruption. This phase of the eruption was not covered in 1986 as no observations were available between 6 and 11 days after the beginning of the outburst.

4. Conclusions

These results show that it is possible to detect and describe superhumps from visual estimates. It is important, however, that the observations are carried out in a consistent and systematic manner. The estimates should be made by using the Argelander method or the fractional method of N. R. Pogson (1829 - 1891). The results may be improved if several observers cooperate. It also should be mentioned that the superhumps for the brighter SU UMa stars can be detected photoelectically bu using equipment within the reach of amateurs. This is a field where experienced amateurs can make important contributions.

References

Hurst, G.M. 1990, Personal communication.
Mattei, J.A. 1990a, AAVSO Alert Notice No. 123.
Mattei, J.A. 1990b, AAVSO Alert Notice No. 124.
Robinson, E.L., Shafter, A.W., Hill, J.A., Wood, M.A., & Mattei, J.A. 1987, ApJ, 313, 772.
Shafter, A. W., Szkody, P., & Thorstensen, J.R. 1986, ApJ, 308, 765.

CONCLUDING REMARKS

John R. Percy
Erindale Campus
University of Toronto
Mississauga, Ontario
Canada L5L 1C6

This has been a memorable and successful meeting, thanks to the warm hospitality and effectiveness of our Belgian hosts, and to the contributions of the dozens of other people who helped in so many ways. We have met old friends and new ones, from all parts of the world. We have shared in the excitement of "the new Europe", especially with the presence of so many colleagues from eastern Europe. The benefits of international cooperation in science have never been as obvious as they have been this week.

Amateur-professional cooperation has also been a prominent feature of this meeting. The review papers have provided numerous examples of how amateurs contribute to variable star astronomy, and it is not surprising that the demand for AAVSO data and services has increased by a factor of ten in the last two decades. It is remarkable how the "small science" of visual observations can support the "big science" of space astronomy, large ground-based telescopes, and modern astrophysical theory.

What about the future? We can expect that the contributions of amateurs to astronomy will continue, but only if we give careful thought to the way in which astronomy is evolving, and how we can adapt. One of the useful functions of this meeting was to provide an opportunity to discuss some of these issues, and to begin to formulate some strategies. The following comments reflect my own opinions, and not necessarily those of the AAVSO staff or council!

1. New Technology

It is easy to imagine that new technology will soon leave the amateur behind, but recent history does not support that belief. Until about 1980, photoelectric photometry was specialized and expensive, and only a handful of amateurs were involved in this field. They were usually individuals with interest and skill in electronics. Indeed, one of the benefits of amateur involvement in astronomy is that they bring their particular talents from their own vocation. By 1980, simple, relatively inexpensive photometers became commercially available. Through the efforts of enthusiastic amateurs, sympathetic professionals, and photometer manufacturers, the field was revolutionized. The AAVSO's sister organization IAPPP (International Amateur-Professional Photoelectric Photometry) was both a result and a contributor to this revolution.

CCD cameras - the next revolutionary technology - are already upon us. Both the detectors and the computer hardware and software to operate them are rapidly decreasing in price. Many amateurs have the necessary skill, enthusiasm and resources to join the revolution. They, the professionals, and the manufacturers must again join forces to make sure that suitable scientific projects are identified, and that instruction and motivation are provided. Amateurs may need access to some resources, such as software and electronic information networks. It may be desirable to establish a new organization or publication, or to appropriate the use of an existing one.

Robotic telescopes is another technology which may have a significant impact on variable star astronomy of the future. Will it make amateurs obsolete? Not likely, for many reasons. Amateurs were instrumental in developing robotic telescopes, and should and will continue to advance the technology. There are some observing projects which are unsuited to the robotic approach; human observers can concentrate on these. Human observers can pass on their enthusiasm to others, either individually or in groups like the AAVSO, and hence recruit more amateurs to either observing or to technology development. Humans can progress from observing to analyzing and interpreting data, whether obtained by human observers or by robots. Science is a human endeavour: amateur astronomy satisfies a human need for activity and involvement, and is an antidote to undue emphasis on computers and automation. Observing can be done at little or no cost, wherever there is a telescope. Besides, there are more than enough stars for human and robotic observers alike!

2. Archiving Data

The AAVSO now has the largest archive of visual observations of variable stars, primarily because it is international in scope. Variable star observers and organizations from around the world submit their observations to this archive. This is very effective for the astronomers who wish to use the data, since they need to contact one organization only. One problem is that, before the data can be given to the user, it must be evaluated and edited. In contrast to the data in many astronomical archives, AAVSO visual observations come from thousands of observers - each with their own telescope, and their own unique physiological "detector". Evaluation is essential but time-consuming. The AAVSO (and its sister organizations) must find ways of simplifying the evaluation process, so that data is quickly and easily available. AAVSO headquarters staff are currently developing computer-based techniques which makes this possible.

Another problem is that of advertising the content of the archive. To some extent, this can now be done by circulating the "validation file", a list of all the stars on the AAVSO visual observing program, together with their class, range, period and spectral type. The potential user needs to know, however, how extensive the observations are. This should not be difficult, once the entire AAVSO database is in electronic form. The computer can keep track of how many observations are available for each star in each year, and thus provide a comprehensive validation file on-line. This would reduce the number of time-consuming personal contacts between AAVSO headquarters staff and potential users. NASA is presently compiling an astrophysical data directory of information on multi-wavelength observations of stars, and the AAVSO has contributed information to this database.

Finally, there is the question of how visual observations should be published. In the past, the AAVSO attempted to publish all observations as light curves, in "reports" issued every few years. This becomes expensive and time-consuming when millions of observations of thousands of stars are being archived each decade. Observers do like to see their observations "in print", but this mode of publication is rapidly becoming a thing of the past. It is more practical and cost-effective to archive the data electronically, and make it available (preferably electronically) only to those who really want it or need it. Now that the AAVSO can produce computer-generated light curves almost instantly, those observers who absolutely must see their observations in print can be supplied with light curves of some of their favourite stars (and several dozen have already taken advantage of this possibility), while most observers can be content to know that their observations are in the AAVSO computer, in the safest, cheapest, most effective form.

3. Feedback between Observers and Users

Publication of observations is one form of feedback to observers. They can see the results of their observations, and how they compare with observations made by others. But that is rather superficial feedback. What observers need is more concrete information about the use and value of their observations. This can best be done through presentations by users to observers, either in print or (even better) face-to-face. That is one reason why this meeting, and this book, have been worthwhile. I believe that the AAVSO and its observers and users should constantly be sponsoring, supporting and attending meetings, international, national and local. The results of amateurs' observations should be featured in astronomical books and periodicals. Users should constantly be reminding their professional colleagues of the value of such observations. It is the least which we can do.

4. Publication of Results of Observations

How should the results of visual observations of variable stars be published? This is one aspect of the topics discussed above. The closest thing to an international journal of variable star research is the International Astronomical Union's *Information Bulletin on Variable Stars*, but the IBVS does not accept papers based on visual observations, and is also under financial pressure to reduce its circulation (it is presently free) and its volume of material. Many national variable star observing organizations publish their own journals, but most of these have very limited circulation, and are not covered by major abstracting services such as *Astronomy and Astrophysics Abstracts*. The *Journal of the AAVSO* circulates fairly widely, and is refereed and abstracted; it welcomes original research papers on all aspects of variable star astronomy, including those based on visual observations. The most desirable solution would be to publish the results in regular astronomical research journals, but these are under pressure to be selective, and might only publish the briefest, most significant papers.

I believe that the most important consideration is that papers be <u>abstracted</u> and <u>available</u>. Bibliographic databases such as SIMBAD (maintained at the Stellar Data Centre at the Strasbourg Observatory) play an increasingly important role in variable star astronomy. Every significant fact and paper about a variable star should be in these databases. As for accessability: it is not necessary for every journal of every national variable star observing organization to be in every library, but they should be in at least one central location (such as the AAVSO library), where they can be photocopied, at cost, for potential users.

5. Coordination of Observing Programs

As a user of AAVSO visual observations, I am constantly aware that some stars seem to be over-observed, others hardly observed at all. This frustrates my sense of organization, but I know that it is difficult to tell observers (especially worldwide) which stars they should observe and which ones they should not. After all, many observers are constrained to observe stars of particular brightness and range. John Bortle does an exceptional service to the AAVSO by editing the *AAVSO Bulletin*, which gives an overview of the behaviour of the more popular variables, and includes the "director's request" for more coverage of those variables which need it.

An obvious but helpful approach was raised at this meeting, symbolized by a graph of apparent brightness *versus* range. This graph can be divided into four quadrants: bright and faint, large-amplitude and small-amplitude. What we need to do is to start observers in the bright, large-amplitude quadrant and, when they are able and confident, move them into the other three quadrants by whatever means we can!

6. Attracting New Observers

The AAVSO has recently become concerned about recruiting a new generation of observers. Many of the best and most active observers are middle-aged or older. Where are the younger ones? This problem does not seem to be universal. It is apparent from this meeting that there are many capable young observers in Europe and elsewhere. What is the difference? It seems to me that there is a social element to variable star observing in these countries, whereas much variable star observing in the US is done by individuals working alone. Occasionally, more experienced observers act as mentors for younger ones; this should be encouraged. Observers should give presentations and demonstrations at meetings of astronomy clubs, and especially at "star parties". The AAVSO is presently embarking on an important new initiative which may attract thousands of young observers: we are applying to various agencies for funds to develop a flexible, open-ended set of hands-on activities and projects on variable star observing, for use in science and math courses in schools, and eventually in universities. In this way, we hope to "turn on" the younger generation to the thrill of astronomical observation and discovery. This same material will be available to amateur astronomers in general, to help them get involved in this field.

7. International Cooperation and Coordination

This meeting has achieved its goal of recognizing and encouraging international cooperation and coordination in variable star observing. But how is this to be continued and expanded? An International Union of Amateur Astronomers existed for many years, but was never very active or effective. Amateur astronomy now has official support from the International Astronomical Union. At the 1988 IAU General Assembly in Baltimore, the following resolution was passed: "The 20th General Assembly of the IAU, *recognizing* the long-standing tradition of excellent and practical collaboration which has existed between amateur and professional astronomers, particularly during the first seven decades of our Union's existence; *noting* that additional communication for common projects is needed today between amateurs and professionals; *recommends* that a Working Group be established to foster this cooperation; and *instructs* the General Secretary to communicate this proposal to the Executive Committee, and to arrange for publication of this proposal by national and international organizations, both amateur and professional". The Working Group has been established, and is actively planning for the future; it will be interesting to see what develops. Key amateurs who are not IAU members could be elected as "consultant" members of IAU Commission 27; a similar arrangement exists in Commission 46 (The Teaching of Astronomy).

In the field of variable star observing, I think it would be useful to set up our own international coordinating group, consisting of representatives from national organizations such as those at this meeting, plus a dozen or so professionals with an interest in amateur-professional cooperation, and expertise on different types of variables. The representatives should exchange newsletters and reports, so that each is aware of the activities of the other organizations, and can cooperate or collaborate as necessary. I do not believe that we should establish a cumbersome administrative structure, or a new publication. Cooperation is best accomplished by a small group of knowledgeable, enthusiastic people, who are personally acquainted (if possible) and who can act as links to their own organizations. This meeting has laid the groundwork for such a group.

AUTHOR INDEX

Baruch, J., 109-116
Blaauw, A., 3-7
Breger, M., 50-51, 171-184
Busarello, G., 126-134

Cuypers, J., 148-155

Drew, J., 302-310

Evans, R., 88-92

Figer, A., 185-189
Fleet, R., 311-313
Fontaine, B., 221-230

Granslo, B.H., 314-320

Hall, D.S., 95-108
Hazen, M.L., 185-189

Isles, J.E., 231-241

Karovska, M., 255-258
Kellomaki, A., 156-158
Kilkenny, D., 205-213
Klotz, A., 221-230

La Dous, C., 279-289
Lampens, P., 60-63
Le Contel, J.M., 117-121
Lipunova, N.A., 55-59
Lloyd, C., 242-246
Longo, G., 126-134

Makela, V., 156-158
Manfroid, J., 75-87
Mattei, J.A., 36-49, 67-74
Mennessier, M.-O., 247-254
Mikolajewska, J., 267-278

Percy, J.R., 11-20, 137-147, 317-322
Pollacco, D., 214-217
Poretti, E., 190-193
Proust, D., 259-264

Querci, F.R., 221-230
Querci, M., 221-230

Samus, N., 52-54
Shore, S.N., 290-301
Sterken, C., 75-87, 126-134, 190-193

Verdenet, M., 259-264
Viotti, R., 194-204

Walker, E.N., 117-121, 122-125
Williams, T.R., 21-35
Wisniewski, W.Z., 159-168

SUBJECT INDEX

This is a brief index. Individual stars (listed under that heading) are included only if they are discussed in the book in some detail. Individual astronomers are mentioned on pages 11-20 (primarily professionals) and 21-35 (primarily amateurs). Major or defining references are in bold face.

alpha^2 CVn stars, **145**
amateur astronomy, 3, 14-15, 21-35, 120-121, 228, 287-288, 300, 310
American Association of Variable Star Observers, 15, 21, 21-35 <u>passim</u>, **36-49**, 67-74, 100, 233, 249-253, 257, 286-287, 322-324
archives, 50, 322-323
asteroids, 101, **159-168**, 258

Be stars, 144
Beta Cephei stars, 117-118
binary stars (see also cataclysmic and eclipsing variables), 60-63, 257, 259-264
BL Boo stars, 57
BL Herculis stars, 143
British Astronomical Association, 15, 73, 231-241

Carte du Ciel, 4
cataclysmic variables (see also U Gem, Z Cam, SU UMa, novae), 68-72, **267-319**
CCD's, 18, 88, 110, 113-114, **126-134**, 161, 167, 185, 322
Cepheids, **38**, 55-59, 100, 138-143, 147, 190-193
chaos in variable stars, 143, 223, 230, 240
classification of variable stars, 16-17
comets, 101, 165-166
computers, 19
computer networks, **156-158**
coordination of observing programs, 67-74, 171-184, 324

Delta Scuti Stars, 100, 117, 119, 138, 140, **171-184**

eclipsing variables, 17, **40**, 100, 102-103, 144
European Southern Observatory, 7, 119

FU Orionis stars, **145**

galaxies, 55-59, 88-92, 126-134
General Catalogue of Variable Stars, 50, **52-54**, 55-59, 233

HIPPARCOS, 60, 71, **249-253**
history, 11-20, 21-35
Hubble-Sandage Variables, **194-204**

Information Bulletin on Variable Stars, 50, 323
infrared astronomy, 18-19, 205-209, 258, 270
International Amateur-Professional Photoelectric Photometry (IAPPP), 100, 321
International Astronomical Union, 6, 50, 51, 157, 325
irregular red variables, **223**

luminous blue variables, 194-204

Magellanic Clouds, 55-59, 138-139, 188, 202, 204
mass loss, 141-143, 255, 302-310
mid-B variables, 117
Mira stars, 13, **38**, 44, 100, 118, 142, **221-230**, 231-241, 242-246, 247-251, 255-258, 259-264

nomenclature of variable stars, 37
novae, **39**, 45, 100, 290-301

period analysis, **148-155, 179-183**, 191, 223, 247-251
period changes, 102-103, 141, 150-152, 206, 231-241, 242-246
photoelectric photometry, 17-18, 85, **95-108**, 117-121, 122-125, 171-184, 227-228, 230, 288-289, 321
photography, 15-16, 126, **185-189**
plan of selected areas, 5
polarimetry, 256-257
pulsating variables, 17, 38, 100, **138-144**

radio astronomy, 18, 267, 273
R Coronae Borealis stars, **40**, 49, **205-213**, 214-217
recurrent novae, **39**, 46, 100, 291
robotic telescopes, **109-116**, 322
rotating variables, 17, **40**, 100, 103-104, 144-145
RR Lyrae stars, 38, 100, 139, 142, **187-188**
RV Tauri stars, 38, 44, 100, 143

semiregular variables, **38**, 45, **222-224**
spectroscopy, 16, 226, 260, 267-278
stars, evolution, 137-147, 255-256
stars, individual
    DX And, 309-310
    AG Car, 196-199
    eta Car, 142, 199-201
    T Cep, 152, 155, 237
    omicron Cet, 256-257
    TT Cra, 311-313
    CH Cyg, 275-276
    P Cyg, 201
    IR Gem, 308-309
    X Oph, 259-264
    EW Sct, 190-193
    SW UMa, 314-319
    alpha UMi, 141
sun, 140, 145

supernovae, 12-13, 25, **39**, **88-92**, 100, 129-130, 146, 202, 299
supernovae, visual observations, **88-92**, 146
SU UMa stars, **39**, 280-281, 285
symbiotic stars, **39**, 48, 100, **267-278**

T Tauri stars, **145**

U Geminorum stars, **39**, 46, 47, 279-289, 302-310, 311-313, 314-319
ultraviolet astronomy, 19, 68-72, 257, 270-271, 283, 295-297, 302-310

vision, 76-79+
visual observation, 40-42, **67-74**, **75-87**, 88-92, 190-193, 311-313, 314-319

white dwarfs, 143-144

X-ray astronomy, 19, 68-72, 110, 273

Z Camelopardalis stars, **39**, 47, 280-281, 281
ZZ Ceti stars, 118, 143-144